CHOOSING & USING
the Right
MILLING MACHINE

CHOOSING & USING
the Right
MILLING MACHINE

RICHARD REX

INDUSTRIAL PRESS, INC.

Industrial Press, Inc.

32 Haviland Street, Suite 3
South Norwalk, Connecticut 06854
Phone: 203-956-5593
Toll-Free in USA: 888-528-7852
Fax: 203-354-9391
Email: info@industrialpress.com

Author: Richard Rex
Title: Choosing & Using the Right Milling Machine
Library of Congress Control Number: 2022936386

© by Industrial Press, Inc.
All rights reserved. Published in 2023.
Printed in the United States of America.

ISBN (print):   978-0-8311-3684-0
ISBN (ePUB):   978-0-8311-9619-6
ISBN (eMOBI): 978-0-8311-9620-2
ISBN (ePDF):   978-0-8311-9618-9

Publisher/Editorial Director: Judy Bass
Copy Editor: Judy Duguid
Compositor: Patricia Wallenburg, TypeWriting
Proofreader: David Johnstone
Indexer: Arc Indexing, Inc.

industrialpress.com
ebooks.industrialpress.com
1 2 3 4 5 6 7 8 9 10

Contents

CONTENTS

Acknowledgments

The following photos were reproduced with kind permission:

- Little Machine Shop, Pasadena CA: Figure 1-1
- Quality Machine tools (Precision Matthews), Coraopolis, PA: Figures 1-2 through 1-6, 1-13 through 1-17, 1-23, 1-24, 1-26, 1-27 through 1-29, 1-31 through 1-33: Figures 2-1 through 2-7, 2-9: Figure 3-16

Special thanks go to the following, who provided "above and beyond" help and advice: Brad Bacon, for years my go-to source for sensible answers to (literally) hundreds of machining questions, also Ed Bindon and Chris Cleal, who were kind enough to read and clarify many sections of the text.

I am also thankful for the help and encouragement from Judy Bass (Industrial Press) and her colleagues, especially Judy Duguid and Patricia Wallenburg.

Richard Rex

Hendersonville, NC

Introduction

There was a time, say 75 years ago, when a model shopper would speak proudly of his metal lathe as the centerpiece of his model engineering capability. It would likely have at least an independent 4-jaw chuck and a tailstock chuck for drilling and tapping, and be able to cut screw threads predictably. The lathe might even be adaptable for light milling operations, by attaching a vertical slide with vise to the cross-slide.

In model shops today you will usually find both a lathe and a vertical milling machine, which might be as small as a 300-lb benchtop "mill drill" or as large as a one-ton Bridgeport-style "professional" mill.

If you are new to all this, a definition of lathe work versus mill work might be helpful. Both have to do with removing material from a block of metal—the "workpiece." The result will be a functional element that may stand on its own or be part of an assembly. The key difference between lathe and mill is in how the workpiece is handled. On a lathe, the workpiece rotates, and it is cut away by a knife tool moving along the axis of rotation or at right angles to it. Typical products of lathe work are "turned parts" such as spindles, bearings, screws, washers, and circular blanks for gears.

On a milling machine, the *cutter* rotates. The workpiece is clamped to a rigid table, which can be moved left to right (X axis) and front to back (Y axis), in precise increments. The Z axis is represented by the headstock, which can be moved up and down on a vertical column. A motor on the headstock drives the cutter, which is most often a cylinder-shaped end mill, with cutting edges on the bottom face and also, helix fashion, on the outer surface. The end mill removes metal from any exposed surface of the workpiece, resulting in flat, cleanly cut sides parallel to the X, Y, and Z axes. Other cutters often used on the mill are slitting saws (miniature circular saw blades) and drill bits. For drilling operations, the headstock comes with a pinion-driven quill that functions in exactly the same way as a drill press.

Typical mill products are flat-surfaced blocks of metal, often with all six sides at right angles to each other (think "cube"), sometimes drilled for spindles or dowel pins, often tapped for screws.

The lathe and the mill are amazingly flexible machines, but neither is capable of doing useful work right out of the box: Both call for a number of work-holding accessories—chucks, vises, and clamps—and a selection of cutting tools, drills, reamers, end mills, and so on. This book, and its companion book on the lathe *(Choosing & Using the Right Metal Shop Lathe)*, will introduce you piece by piece to the add-ons that get you operational with the least delay and expense.

The mill drill is a fairly recent innovation. Starting in the late 1990s, the first mill drills—all from Taiwan and China—were beefed-up drill presses with vertically adjustable headstocks, and with tables that could be driven in the X (left-to-right) and Y (front-to-back) axes by lead screws with handwheels. The key difference between the earliest mill drills and those sold today is that their round drill press–style columns have mostly been replaced by more rigid dovetailed columns that allow the headstock to be raised and lowered without affecting spindle position relative to the workpiece (a significant shortcoming of the round column type).

Aside from its greater weight and rigidity, the Bridgeport type of mill is quite different in several ways. In the first place, its headstock is not vertically adjustable; instead, the table and workpiece are raised to meet the cutting tool. Second, the table is supported on a massive casting—the "knee"—that runs up and down on dovetailed ways machined into the base, hence the general descriptor for this type of machine as a "knee mill."

Knee mills like this have been a fixture in practically all general-purpose machine shops for the past 75 years. The many variants of the Bridgeport design have been widely copied, with varied success, by many manufacturers in Pacific Rim countries. This has resulted in dozens of similar-looking machines on the market.

When choosing a mill, one thing you will notice as you survey the catalogs is that Asian mills of a given size tend to have similar features. They might just be more than similar, even identical, thinly disguised by the importer's paint job and label. But that's not always the case. Some distributors call for special features and quality control that won't be found on other machines. Also bear in mind that there are dozens of manufacturers making what seems to be the same product, but that doesn't mean that parts will be interchangeable, or of uniform quality. There is literally no standardization, only the similarity that comes, certainly in the case of knee mills, from having copied Bridgeport machines over the years.

Unlike lathes, which have not changed fundamentally in the past 100 years, vertical mills—especially bench mills—are a recent addition to the small model shop. This may account for the relatively small amount of published how-to material—aside, that is, from the thousands of videos on the internet. However, take care: Some are misleading, and a few are just wrong. On the plus side is Tubal Cain, a one-time machine shop instructor and now publisher of many beginner-level videos, all of them reliable it seems to me. For more current advice, there are a few thoughtful machinists making helpful videos. Three examples (2021): Stefan Gotteswinter, This Old Tony, and Joe Pieczynski. Those new to mill work should find this book helpful when working on any size and type of vertical mill. Experienced users may also find useful reminders and new material.

Another publication it's good to have on hand is *Machinery's Handbook*, now in its 31st edition, available in print and on CD. It is recognized throughout the United States as *the* reference for every conceivable machining question. With its almost 3,000 pages of fine print, *Machinery's Handbook* is perhaps over the top for anyone starting out, but there may come a time when you really do need the "official word."

Examples of Frequent Mill Operations

Fly cutting a surface.

Precise drilling.

Milling a slot.

End milling a top surface.

Milling a dovetail slot.

Milling a shoulder.

Using a slitting saw.   Squaring off stock.

INCHES TO MILLIMETERS CONVERSION

Most machine tools today come from countries using the metric system. Additionally, many projects call for metric measurements and metric hardware. To convert from inches to metric, and vice versa, all you need remember is the number **25** as a substitute for the "real" 25.4.

inches to mm

Multiply by **100**, then **divide** the result by **4**   **x 100 ÷ 4**
Example: 1" is approximately 100/4 = 25 mm

mm to inches

Multiply by **4**, then **divide** the result by **100**   **x 4 ÷ 100**
Example: 8 mm is approximately 32/100 = 0.32

In this book, the word "mil" is sometimes used to signify 1/1000" (one-thousandth of an inch).

CHAPTER 1

Choosing a Milling Machine

CONTENTS AT A GLANCE

1-1 A SELECTION OF MILLING MACHINES

Figures 1-1 through 1-6 show a selection of milling machines. *Machines similar in appearance to those in the figures are in some cases available from more than one supplier, but their overall quality and specifications may differ.* All the machines shown in this section are usually offered with R8 spindles.

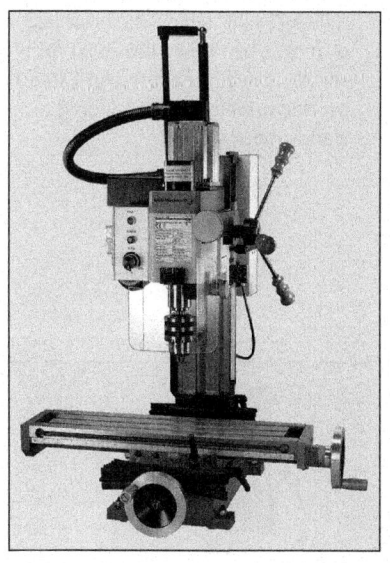

FIGURE 1-1 Small bench mill. A 7" x 17" table, 0.7-hp brushless dc motor, belt drive, 100–2500-rpm spindle, weight about 170 lbs.

FIGURE 1-2 Large bench mill. A 7" x 27" table, 1-hp brushless dc motor, belt drive, 50–2500-rpm spindle, weight about 275 lbs.

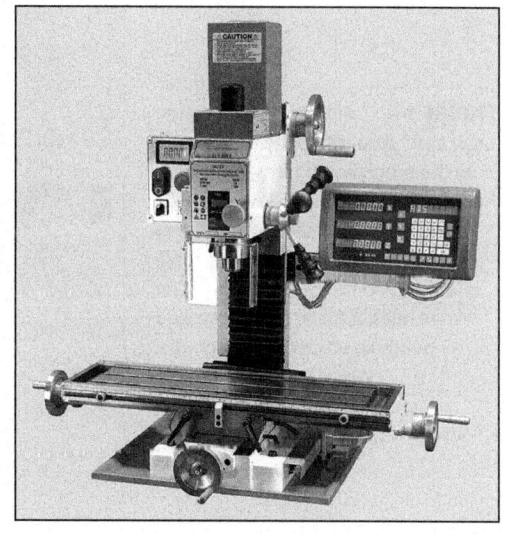

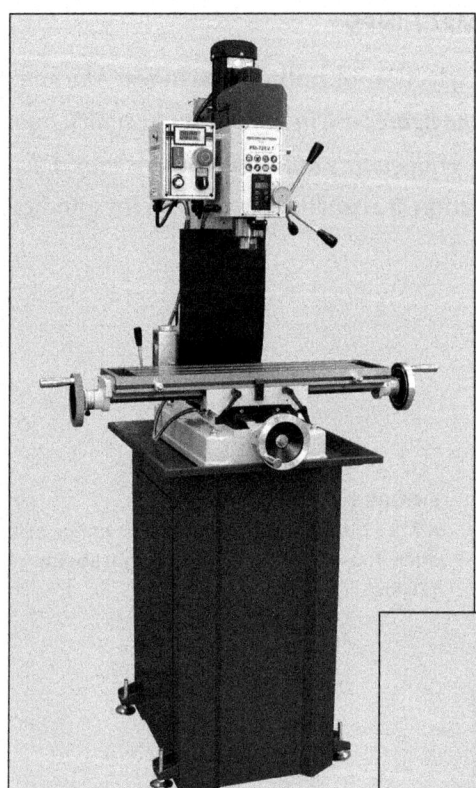

FIGURE 1-3 Intermediate bench mill. A 7" x 28" table, 1-hp brushless dc motor, belt drive, 100–4000-rpm spindle. Usually mounted on a stand as shown. Weight, not including stand, about 400 lbs.

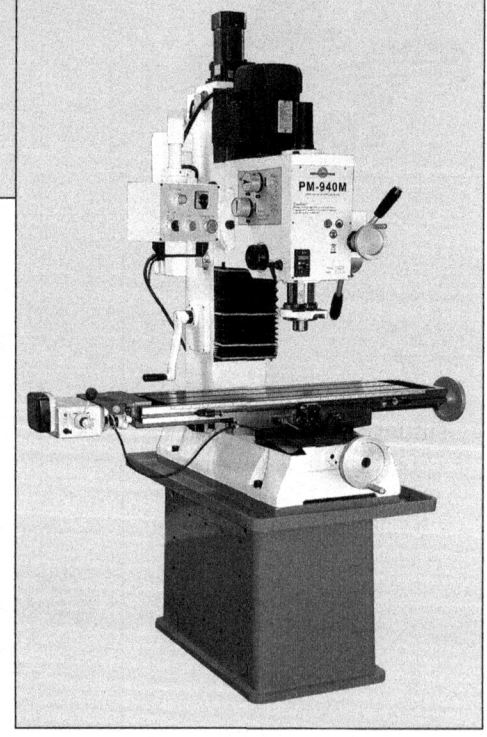

FIGURE 1-4 Large bench mill. Same basic configuration as the bench mills in the preceding figures—movable headstock, not movable knee. Usually supplied with a cast-iron stand, as shown. Measuring 9" x 40", its table is larger than many Bridgeport-style knee mills. 2-hp ac motor, 6-speed gearbox, 90–2000 rpm spindle. Weight, including stand, about1,300 lbs.

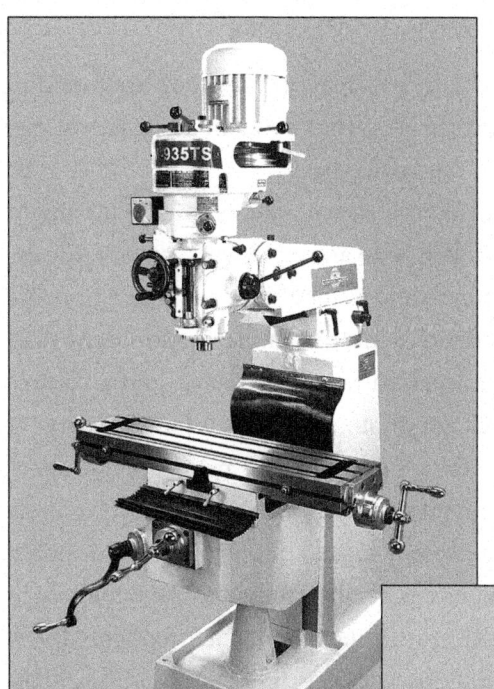

FIGURE 1-5 A series of Bridgeport-style knee mills with Vee-belt drive, table sizes 9" x 35" to 10" x 54", 3-hp ac motor, 4-speed Vee-belt drive with back gear, 80–2700-rpm spindle, weight 1,500–3,000 lbs.

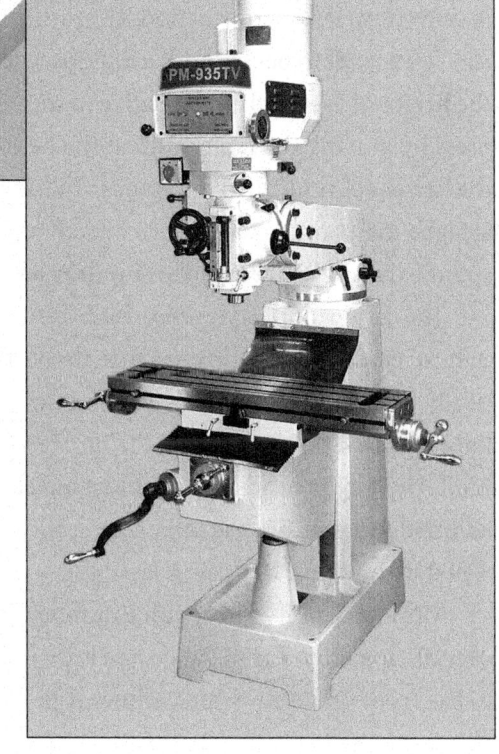

FIGURE 1-6 Knee mill series similar to Figure 1-5, but with a mechanically variable flat-belt drive, 70–4,000-rpm spindle, weight 1,500–3,000 lbs.

1-2 BENCH MILL VERSUS KNEE MILL

Vertical milling machines of the sort described in this book are found in almost all model shops and industrial labs. The two main categories of vertical mills are the bench type and the knee type.

Sometimes called a mill drill, the *bench mill* is what you might expect—a machine that can be bolted onto any rigid work surface, ideally 30" or more above the floor. Stands, some welded steel, some cast iron, are available for most bench mills. Bench mills are similar in concept to the standard drill press. Both have a work surface—the table—and a vertical column that supports a headstock and motor, but that's it for similarity.

The main differences between a mill and drill press are the mill's heavier, more robust construction and the way the workpiece is handled. On a drill press, the workpiece is usually moved into position by sliding it on the table. On a mill, the workpiece is firmly clamped to the table, which is moved by lead-screw action in precise increments, left to right (the *X axis*) and front to back (the *Y axis*). Another key difference is in the *Z axis*, the height of the headstock relative to the table. This is adjusted on the drill press by moving the table, and on the bench mill it is adjusted by moving the headstock. One final difference: The drill press spindle is always exactly at right angles to the table. On a bench mill, the headstock (and with it, the spindle) can be rotated and set up to ± 90° left or right relative to the table.

The Bridgeport-style *knee mill* is functionally similar to the bench mill in most respects, except that its headstock cannot be adjusted vertically (it has no column). Z axis motion comes instead from a movable knee, a robust platform for the table, weighing several hundred pounds in itself. Unlike the headstock on most bench mills, knee height can easily be adjusted in small, precise increments, allowing the cutting depth to be fine-tuned within ± 0.0005", even less.

Most knee mills come with a number of other features that add to their overall capability. For instance, the knee mill headstock is usually mounted to the front of a ram, which allows it to be repositioned front to back (in some cases, the ram sits on a turret that allows full-circle rotation in the vertical axis). Also, just as on a bench mill, the knee mill headstock can be rotated and set ("trammed") at ± 90° relative to the table. On all knee mills—but never on bench mills—it can also be rotated forward and backward by a few degrees, in some cases as much as ± 45°.

Finally, perhaps the most obvious distinguishing feature of the knee mill is its weight. Even the smallest of knee mills is significantly heavier than any bench mill. *Why is that important?* The answer can be summed up in one word: "rigidity." This is the key factor that determines maximum depth of cut and, to a lesser extent, smoothness of the cut surface. (That said, all bench mills can do a comparable job, simply by using less aggressive cuts and lower traverse speeds.)

1-3 MACHINE SIZE AND WEIGHT

A mill is usually "measured" by its table size, but that can be misleading. Many bench mills have tables as large as the most popular knee mills. The other key metrics are weight and (of course) purchase price. Prices vary widely from one distributor to another, but here are some 2021 statistics (shipping is usually extra):

Bench Mills

- The smallest bench mill has a 7" x 27" table and weighs about 300 lbs. The largest bench mill has a 9" x 40" table and weighs about 1,000 lbs.
- Bases are available for most stands, which eliminates the need for a dedicated bench. Some stands are cast iron, weighing about 300 lbs.; others are welded metal, about one-third of the weight.
- Prices go from about $1,500 to $4,000. Essential accessories cost about $500.

Knee Mills

- Knee mills range in table size from 8" x 35" (1,400 lbs.) to about 10" x 45" (3,000 lbs.). All are floor-standing.
- Prices go from about $4,500 to $6,500. Essential accessories same as for bench mills.

1-4 ADDITIONAL FACTORS TO CONSIDER

Two more factors need to be considered:

- **Ceiling height.** For bench mills, be sure there's enough headroom in the shop for the headstock at its maximum height. Knee mills have no height adjustment. For both bench and knee mills, be sure there is headroom for the hoisting/rolling equipment when installing.
- **Floor loading.** Before installing any milling machine, knee mills in particular, be sure the floor can take the machine weight plus a 500-lb allowance for workpiece and work-holding items.

1-5 MILL VERSUS DRILL PRESS SIMILARITIES

There are similarities between all small vertical milling machines and the traditional drill press, such as the vintage example shown in Figure 1-7.

The main similarities are:

1. A vertically movable *quill* that encloses the spindle.
2. A *drill press lever* that propels the quill downward.
3. A quill *clamp* to lock the quill firmly in position.
4. A variable-speed *spindle drive system.*
5. A *headstock* that can be moved up or down on a vertical column. Note, though, that this feature applies only to bench mills. A key question regarding the column: *Is the column round or dovetail?* Headstock alignment can be a serious

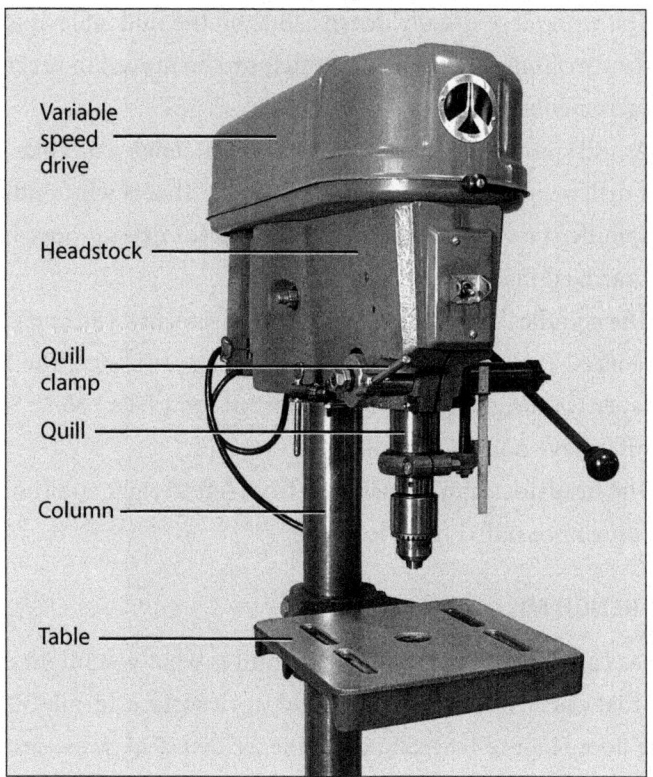

FIGURE 1-7 Drill press.

problem with a round column. Most of today's bench mills have dovetails—much better.

1-6 MILL VERSUS DRILL PRESS DIFFERENCES

Every vertical mill is a *part-time drill press*, but there's more to it than that. Here are the main differences:

1. Massive, rigid construction, a lot more cast iron.
2. Heavy T-slotted *movable table* on dovetail ways, with precise position measurement capability. Optional digital readout (DRO) on left-to-right (X), front-to-back (Y), and up-down (Z) axes.

3. The workpiece usually doesn't slide on the mill table; it is firmly *clamped* to the table, which can be moved in precise increments.

4. A mill spindle is designed for both axial *down load* (like a drill press) and also *side load* (radial). That is why a mill spindle runs in tapered roller bearings (or deep-groove ball bearings) inside the quill.

5. The spindle isn't just for drill chucks—use any *R8-compatible device*—end mill holders, collets, slitting saws, etc. (The R8 taper was originated by Bridgeport in the 1950s.) *Note:* Some mills have non-R8 spindles.

6. The headstock can be swiveled from left to right and (on some knee mills) front to back.

1-7 THE BENCH MILL

Sometimes called a mill drill, the bench mill is what you might expect—a machine that can be bolted onto any rigid work surface, ideally 30" or more above the floor (Figure 1-8). Stands, some welded steel, some cast iron, are available for most bench mills.

Most bench mills have these basic features (Figure 1-9):

1. A *saddle* on dovetailed slides that can be moved in precise increments forward and backward by the *Y axis lead screw*.

2. A flat *table* on a second set of dovetailed slides, on top of the saddle, that can be moved in precise increments left to right by the *X axis lead screw* (X and Y axis dovetails in the saddle are at right angles).

3. Three or more *T-slots* on the table for clamping the workpiece or vise.

4. A *column* at the back of the machine, right angled to the table in both the X and Y axes, that supports the headstock and allows its vertical position to be adjusted, usually by a hand-

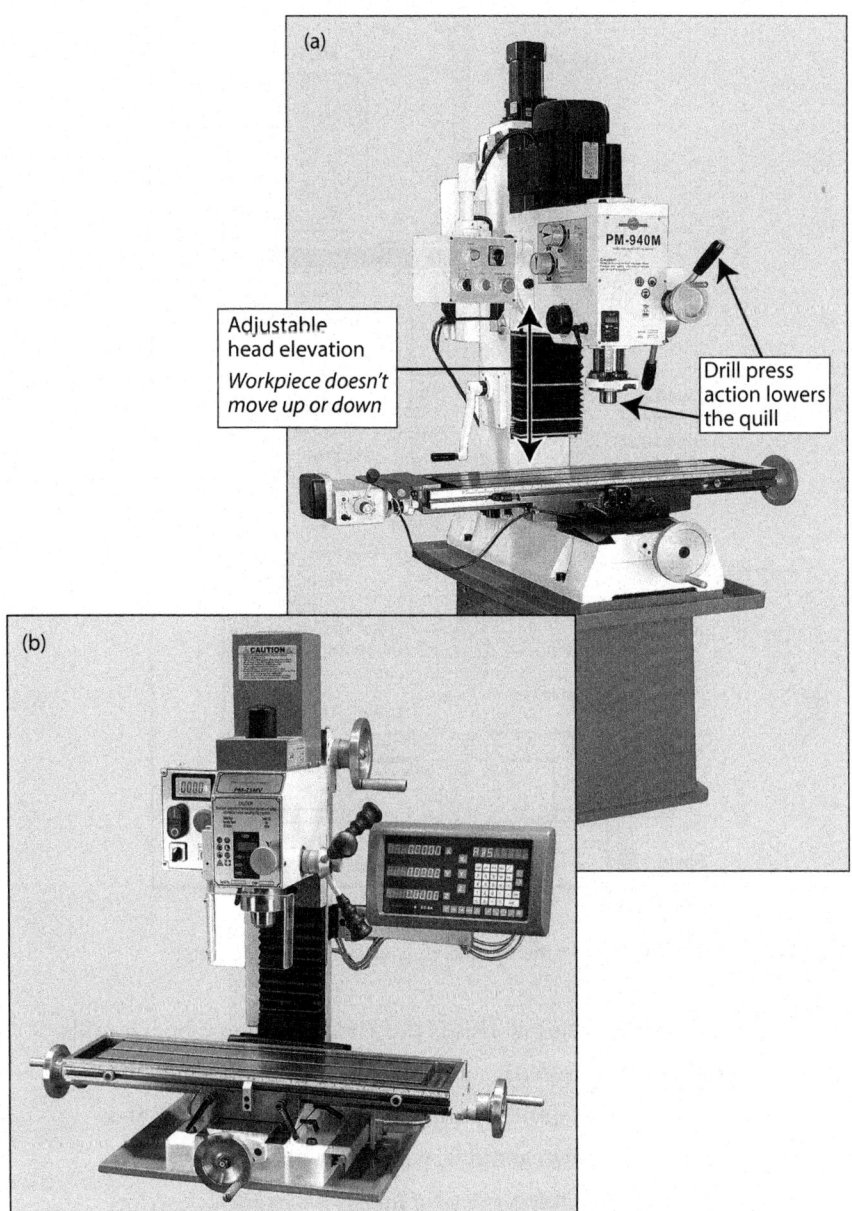

Adjustable head elevation
Workpiece doesn't move up or down

Drill press action lowers the quill

FIGURE 1-8 (a) Bench mill, 9" x 40" table, with cast iron stand. This example includes power feed motors on the X axis (table) and Z axis (headstock). (b) Bench mill, 7" x 27" table. This example includes a DRO unit for precise indication of table and headstock positions.

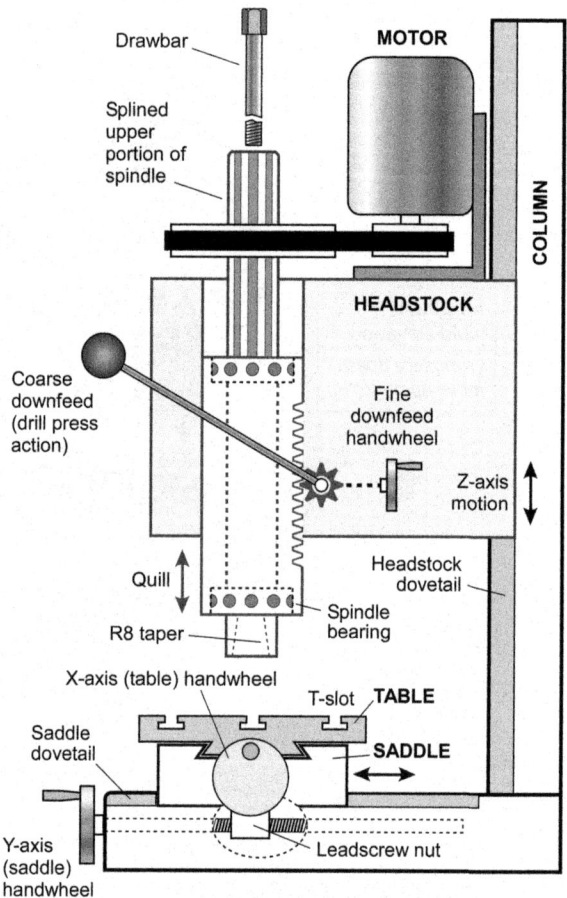

FIGURE1-9 Bench mill schematic. Headstock (Z axis) lead screw omitted.

wheel and lead screw. This is the Z axis (on some bench mills the Z axis is powered).

5. A motor-driven spindle running in a sleeve (the *quill*) that slides up and down about 4" within the headstock casting. *Quill position is independent of the Z axis headstock setting* (unless the quill is locked to the headstock). The quill is controlled by a pinion that engages a rack of teeth on the quill; the pinion is controlled by an external lever, exactly as on a drill

press. Most bench mills have, in addition, a *fine-feed hand-wheel* to adjust the quill position in very small increments.

6. A flared chamber at the bottom end of the spindle for R8 toolholders (examples shown in Figure 1-10). The shanks of all R8 fittings are internally threaded to mate with the draw-bar—see list item #7. (There are—or there used to be—bench mills with Morse taper sockets instead of R8. If you have a choice, R8 is by far the better option.)

7. A drawbar, within the spindle, to secure the R8 toolholder. The lower end of the drawbar is threaded UNF 7/16"-20.

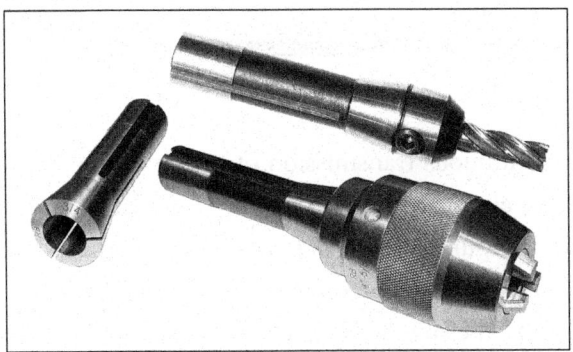

FIGURE 1-10 R8 fittings. *Clockwise, from top:* End mill holder, drill chuck, and collet chuck. Rotation of the fitting before tightening is prevented by the key slot, which engages with a pin or set screw in the spindle bore.

1-8 SPINDLE DRIVE SYSTEM—BENCH MILLS

Just as in a drill press, the mill headstock supports the motor and the spindle drive system. The drive system in all small mills delivers spindle speeds from about 100 to 2000 rpm. If the mill has a single-speed, single-phase ac motor, the spindle drive is usually through a "gear head," a 6-speed gearbox with high and low ranges, 3 speeds in each range.

Some users of bench mills (and lathes) replace their single-phase ac motors with three-phase motors of the same physical size. These motors are powered by a variable frequency drive (VFD), which allows spindle speed to be "steplessly" controlled by a potentiometer. This is easy and elegant, but it still needs the gearbox to ensure adequate torque at the slowest speed.

Another variable-speed system offered by manufacturers of smaller mills is based on a dc motor with a *belt drive*. This, like the VFD just described, also suffers from torque "falloff," which is why belt-drive mills typically come with some form of *intermediate-range selection*, mostly using 2-step pulleys (Figure 1-11).

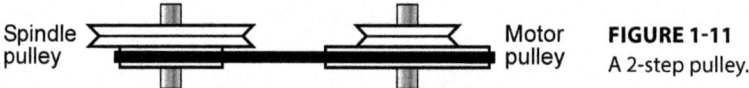

Spindle pulley   Motor pulley   **FIGURE 1-11** A 2-step pulley.

To allow continuous transmission of power as the quill is raised and lowered, the upper portion of the spindle is splined to mate with internal splines on the final pulley or gear of the drive train. (This is functionally the same as the keyed spindle used on many drill presses.)

In most mills, the spindle runs on high-quality *tapered roller bearings* or *deep-groove ball bearings* within the quill.

1-9 TABLE STRUCTURE

All vertical milling machines have a rigid, heavy cast-iron table running left to right (the X axis) on full-width dovetail slots machined into the underside. Figure 1-12 shows a typical cross section. Mating dovetails for the table are on the upper surface of a saddle casting, on the underside of which is another pair of dovetails (the Y axis) at *exactly 90°* to the upper pair seen in Figure 1-13.

Referring to Figures 1-12 and 1-13, the top surface of the table is ground precisely flat to allow the vise and other accessories to be positioned accurately and repeatably in any location. Importantly, too, the T-slots are machined precisely in line with the dovetails on the underside.

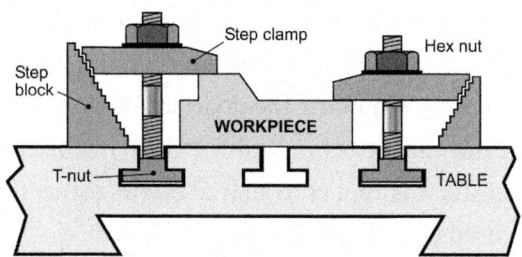

FIGURE 1-12 Cross section of a typical table. Shown here are T-nuts and other "clamping kit" items used to hold down the workpiece. The clamping kit is an essential accessory, typically purchased separately—see Section 4-4 in Chapter 4.

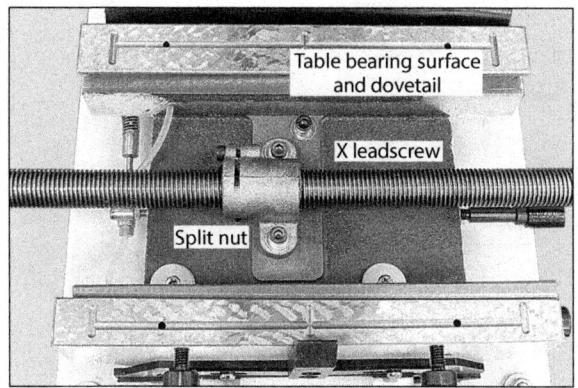

FIGURE 1-13 Table saddle (table removed). This is a massive casting that is moved forward and back by the Y axis lead screw (not shown). Unusual for a small mill, this unit has oil channels for a central "one-shot" lubricating system. The bronze split nut for the X lead screw can be adjusted to minimize backlash.

This, together with flatness, is something you come to rely on instinctively in practically all machining operations.

On almost all mills sold in the United States, the lead screws have 10 TPI (threads per inch), so a full revolution of the handwheel moves the table one-tenth of an inch. The handwheel dials are graduated in 100 divisions, each division being 0.001". The dials are friction-coupled to the lead screws, and can be rotated—without turning the lead screw—to bring any desired graduation in line with a fixed datum pointer (Figure 1-14, inset). This feature is often used to move the table by a precise amount from one location to another. For instance, if you need to drill 0.75" from a first hole, you would zero the dial at the first hole, then turn the handwheel to bring 75 to the datum. "Counting the divisions" is described in Chapter 3, Section 3-5. Meanwhile, bear in mind the matter of backlash, addressed next, in Section 1-10; this applies to all lead screws (other than the recirculating-ball type used in CNC machines).

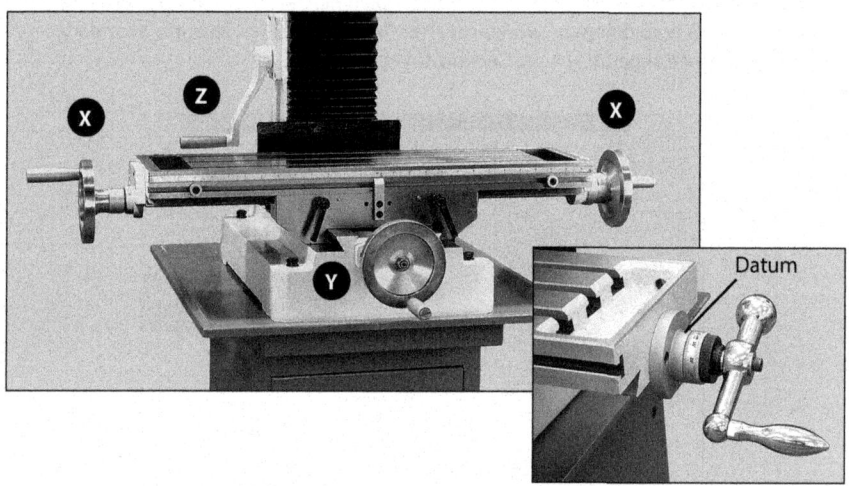

FIGURE 1-14 Bench mill X, Y, and Z controls. The "Z" handle raises and lowers the headstock. On some bench mills, this is near the top of the column. In a few cases, the Z drive is motorized.

1-10 BACKLASH

When alternating between clockwise and counterclockwise rotation of the X or Y lead screw, the handwheel moves freely a few degrees, but the table stays put. The acceptable amount of lost motion—backlash—depends on the user, but ± 0.005" is generally a good compromise. On most mills, there is provision for backlash adjustment, usually a compressible split nut (Figure 1-13).

1-11 OTHER BENCH MILL FEATURES

Unlike the typical drill press, a bench mill headstock can usually be rotated and set at ± 90° side to side relative to the 0° "drilling machine" axis used every day. However, on a bench mill the headstock *cannot be rotated forward and backward.* This does not apply to most knee mills.

1-12 TABLE POWER FEED

On most milling machines, the X axis lead screw can be motorized by an optional power unit (Figure 1-16). Like most deluxe-seeming accessories for the mill, once you have experienced power feeding, you will wonder how you got along without it. Another option like this is the DRO (Section 1-14 below).

1-13 POWER DOWNFEED (QUILL AUTO FEED)

If you're looking for a mill that can handle batch production jobs, but haven't yet purchased the machine, you may want to consider bench

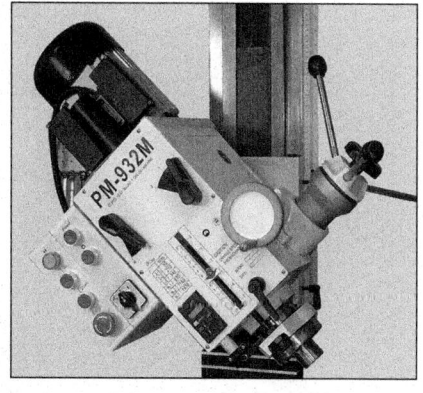

FIGURE 1-15   Mill headstock tilted 45° counterclockwise.

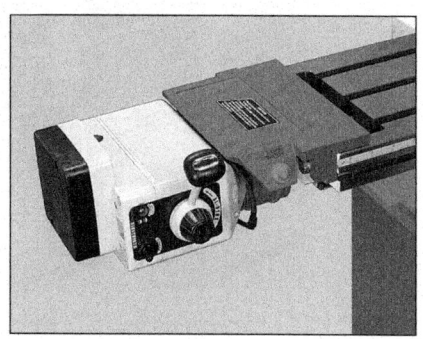

FIGURE 1-16   X axis table power feed.

mills that offer a power downfeed option (this is not usually retrofittable and has to be ordered up front). Power downfeed can automate repetitive, sometimes tedious tasks such as drilling, reaming, honing, and hole boring. (Most knee mills come with auto feed, just as Bridgeport mills have done from the mid-1900s, as described in Section 1-20 later in this chapter).

In Figure 1-17, the down feed levers are pivoted inward to control the quill by hand (drill press style) and outward to move the lever hub and quill under power. Three feed rates are available: 0.1 mm (± 0.004"), 0.18 mm (± 0.007"), and 0.26 mm (± 0.010") per spindle revolution. These

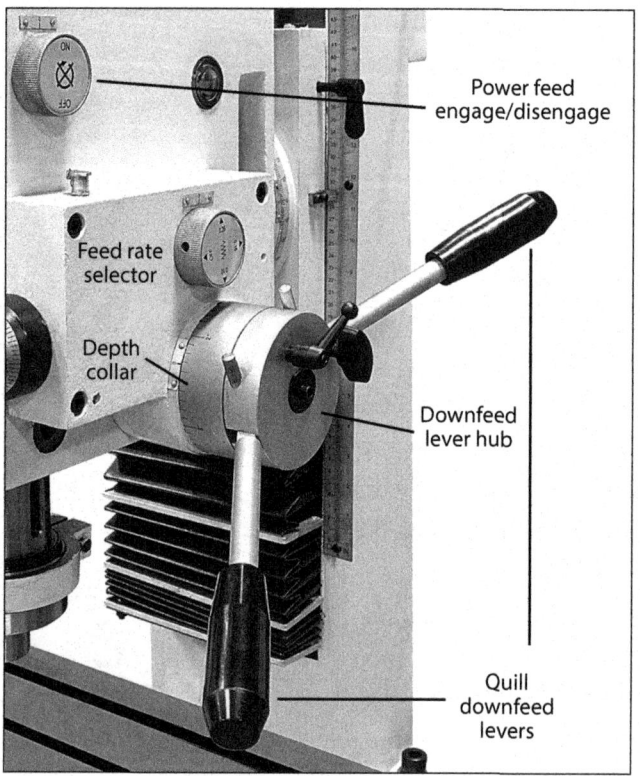

FIGURE 1-17 Power down feed on a bench mill.

numbers, and this hardware configuration, may not apply to all power downfeed add-ons.

In the down feed example in Figure 1-18, a previously drilled hole is being enlarged using a single-point boring tool. One of several ways to do this would be to fully retract the quill (1), then lower and lock the head-stock with the cutting tool just clear of the workpiece. With the cutting tool just inboard of the hole, lower the quill by hand to the desired end point (2); then set the depth collar to zero. Now, with the quill fully retracted again (1), engage the power feed and turn the spindle motor on. If you now pivot the downfeed levers outward, the quill will descend under power until "0" on the depth collar approaches the scale datum—at which point

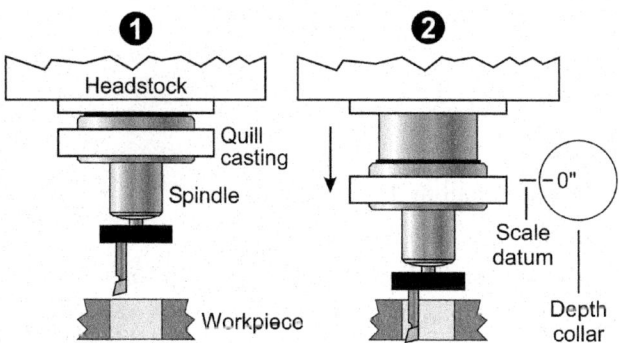

FIGURE 1-18 Boring a hole with power downfeed.

the quill will retract automatically to position (1). Thereafter, the cycle can be repeated as necessary, with the toolholder adjusted each time for a slightly enlarged cutting tool radius.

1-14 DIGITAL READOUTS

A DRO is not essential for precise machining, but it is truly a game changer—a day-versus-night proposition (see Chapter 7). Without a DRO, the mill table is positioned by reference to the graduated dials on the handwheels, in other words by "dead reckoning." This is exactly how it was done, not so long ago, by every machinist everywhere.

Already mentioned in Section 1-10, lead-screw backlash is the main issue with dead reckoning—a full turn of the handwheel may move the table by 0.1", but a full turn in reverse doesn't put the table back where it was. The DRO, on the other hand, reports exact positions. It is completely unaffected by backlash.

Digital readouts have been around for more than 50 years. Unlike earlier models, with super-delicate optical scales and readers, today's DROs for the mass market have more robust scales that allow a fair amount of leeway (magnetic scales can even be cut to length for practically any machine installation). This is made possible by a magnetized ferrite substrate and magneto-resistive sensor instead of the more fragile (but still widely used) optical systems. This has been done without sacrificing resolution or absolute accuracy—magnetic DRO scales can be set up to detect a shift in position as small as 1 micron (about 0.00004").

DROs can be installed on most bench mills (and all knee mills). There are two main versions: 2 axis (XY) and 3-axis (XYZ), shown in Figure 1-19. If a 3-axis DRO is installed, the Z axis displays *headstock* or *knee elevation*; this can be used as a measure of tool depth if (and only if) the quill is locked.

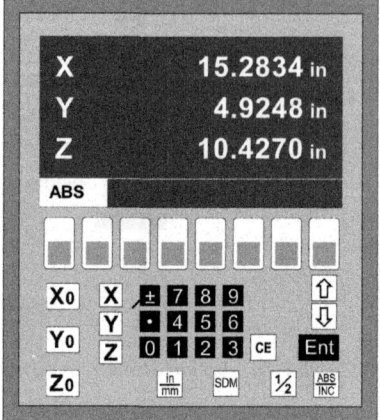

FIGURE 1-19 Typical 3-axis DRO display.

1-15 QUILL DRO

Many bench mills (but not knee mills) have a separate battery-powered DRO for the quill (Figure 1-20). This is *completely independent* of the Z axis DRO scale, which reports headstock elevation on the column. The quill DRO indicates movement of the spindle and is useful in determining precise drilling depth, etc.

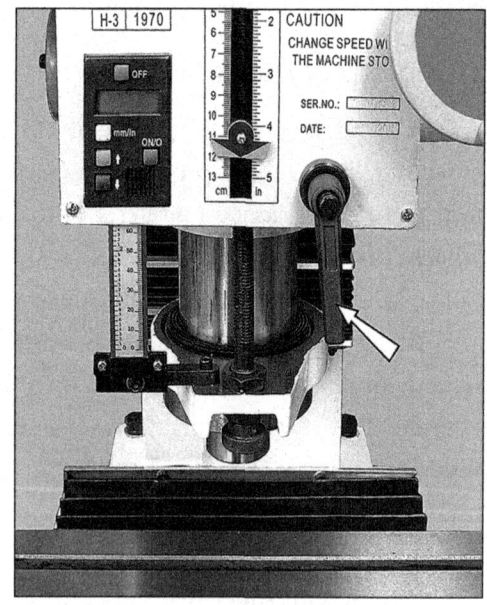

FIGURE 1-20 Typical quill DRO. This is independent of the Z axis DRO (if installed). Unless locked to the headstock by the handle, indicated by the arrow, the quill is positioned by the "drill press" handle or by the fine-feed handwheel. The quill DRO is battery-powered on most bench mills.

BENCH MILL FAQ: ROUND OR SQUARE COLUMN?

Most drill presses, and a few bench mills, have round columns, typically 3" or 4" in diameter. For basic drilling operations, round columns are perfectly functional. However, for milling, they can be troublesome, especially when tool swapping is necessary. One frequent instance of this (there are many) is reaming a just-drilled hole. The headstock must be raised, because the reamer is longer than the drill. *How can that be done without affecting spindle alignment?* With a round column, it is difficult.

Today's solution to the alignment problem is the so-called square column, actually a dovetailed slide. This allows the headstock elevation to be changed by many inches with little effect on the spindle axis relative to the workpiece.

1-16 THE KNEE MILL

The knee mill is functionally similar to the bench mill in most respects, except that its headstock cannot be adjusted vertically.

Z axis motion comes instead from a movable knee, actually a very robust platform for the saddle and table components, weighing several hundred pounds in itself (Figures 1-21 and 1-22). Unlike the headstock on most bench mills, the knee can be raised in small, precise increments, allowing the cutting depth to be fine-tuned within ± 0.0005", even less.

The knee mill headstock is mounted to the front of a ram (Figure 1-21), which allows it to be repositioned front to back. Also, in some cases, the ram sits on a turret that allows full circle rotation in the vertical axis.

Just as on a bench mill, the knee mill headstock can be rotated and set at ± 90° left or right relative to the 0° axis used for ordinary operations. On all knee mills—but never on bench mills—it can also be rotated forward and backward by a few degrees, certainly enough to allow precise

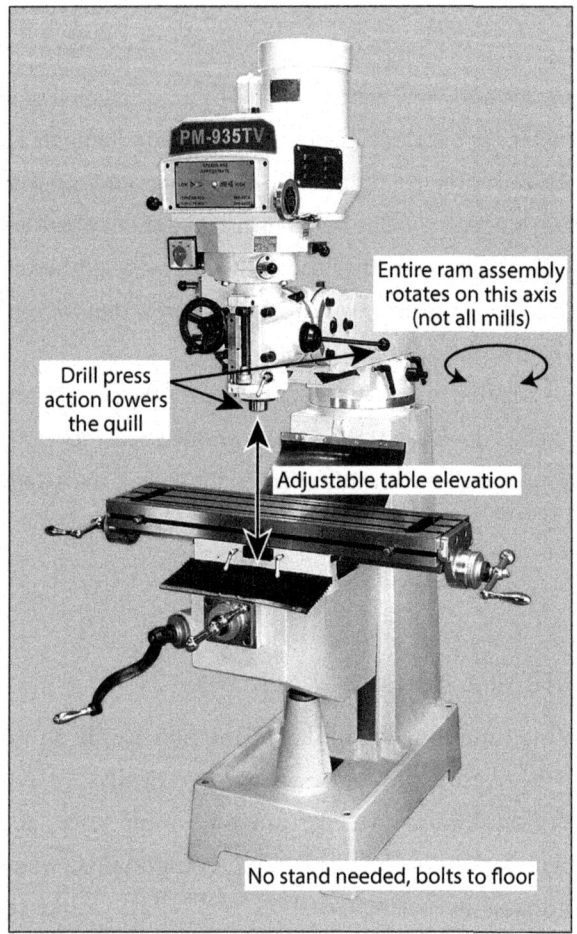

FIGURE 1-21 Typical knee mill. This model has a variable-speed drive system; see also Figures 1-29 and 1-30 later in the chapter.

90° spindle-to-table alignment (known as "tramming"). In some cases, the fore-aft rotatability is as much as ± 45° (Figure 1-23).

Finally, perhaps the most obvious distinguishing feature of the knee mill is its weight. Even the smallest of knee mills is significantly heavier than any bench mill.

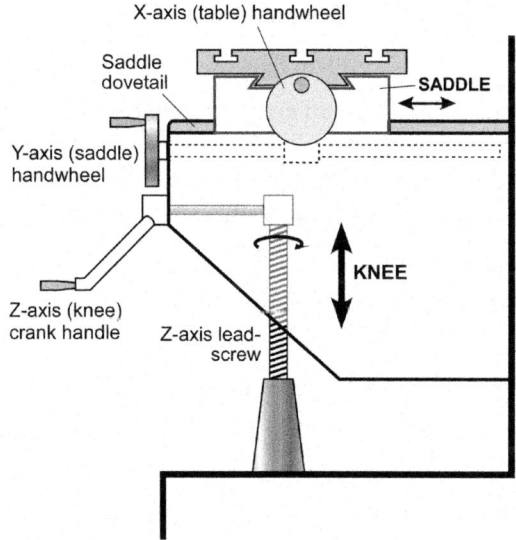

FIGURE 1-22 Knee schematic.

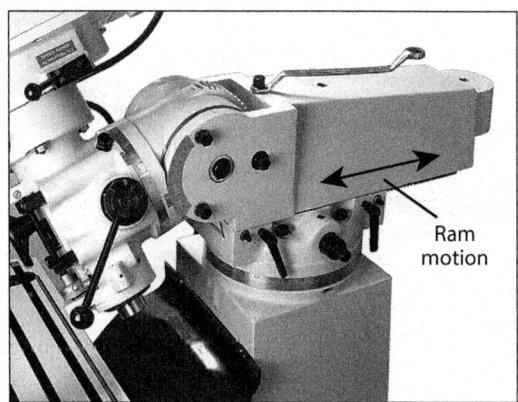

FIGURE 1-23 Knee mill head and ram assembly.

What's important about weight? In a word, "rigidity." This is super-important: It is the key factor that determines maximum depth of cut and, to a lesser extent, smoothness of the cut surface. (That said, all bench mills can do a comparable job, simply by using less aggressive cuts and lower traverse speeds.)

1-17 PLUS FEATURES OF THE KNEE MILL

Positive features of the knee mill include the following. *Note, though, that not all of these features are provided on every knee mill—check with the supplier.*

- Rock-solid, heavy, built-in stand.
- Larger work envelope and greater flexibility than bench mills of similar table size.
- Ram allows front-back positioning of the headstock (in some cases).
- Turret mounting allows ram rotation.
- Rotatable headstock, ± 90° left or right (plus, in some cases, up to ± 45° fore and aft).
- Built-in (usually) power downfeed of the quill.
- One-shot lubrication of machine ways.
- Built-in spindle lock for easier tool changes.
- Power feed option on at least the X axis.

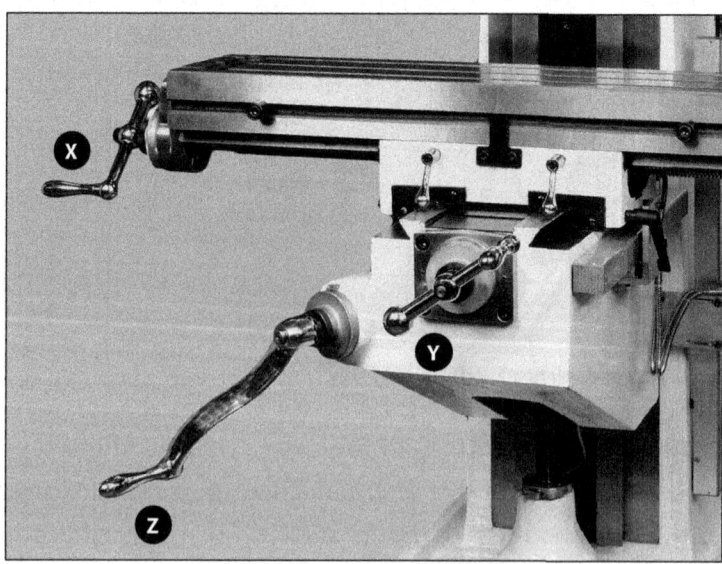

FIGURE 1-24 Knee mill table controls. X moves the table left to right; Y moves the table front to back; Z raises and lowers the table.

1-18 KNEE MILL DRIVE SYSTEMS: 4-SPEED PULLEY

It has been said elsewhere that all vertical mills, whether bench type or knee, are part-time drill presses, so you might expect similarities in the drive train from motor to spindle. This isn't so. The differences are such that experienced users of bench mills can be quite baffled by their first encounter with a Bridgeport-style knee mill—or any of the hundreds of look-alikes out there.

First, a few similarities in the motor-to-spindle train. In almost all cases, the spindle is R8-compatible (see Section 1-7). Additionally, all mills have some means of precisely controlling the quill and locking its position in the headstock. Finally, there is an important common feature that helps ensure usable spindle torque at all speeds—*in both high- and low-speed ranges*. Almost all mills, bench or knee, have three or four choices of speed (or a variable-speed control) in *each range*. However, between bench versus knee there is no similarity at all in the way the speed ranges are engineered. In a bench mill, it can be as simple as a 2-step pulley drive (Section 1-8). It is anything but simple in knee mills—they all use a *back gear* to switch from high range to low range. How this works is shown in Figure 1-25.

The back gear works beautifully, but with two gotchas:

1. It *reverses the drive*, which means that for ordinary milling and drilling, the motor, too, must be reversed—see #6 in Figure 1-26.

2. Switching from the LO range to HI can be a big surprise. *Be sure motor power is OFF.* Move the spindle cam lever to lower the driven pulley; then push on the Vee belt (Figure 1-27) until you *hear the clunk* as the clutch dogs engage—until that happens, *don't run the motor.* Move the back gear crank to OUT or HI (Figure 1-28); then run the motor (normal direction).

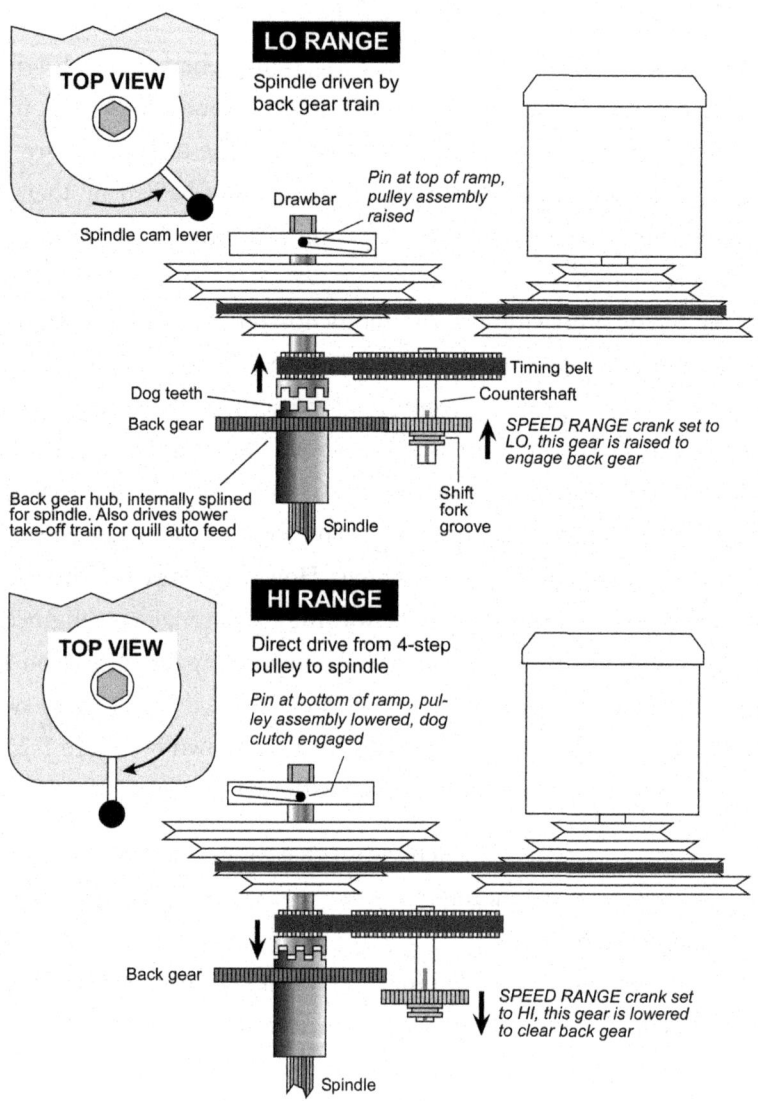

FIGURE 1-25 A 4-step pulley drive with back gear.

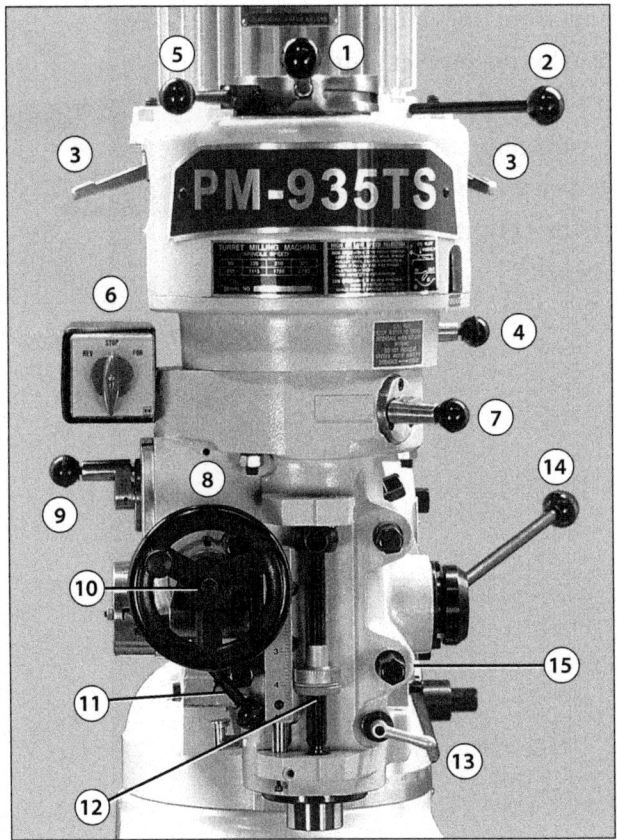

FIGURE 1-26 Typical pulley drive controls, front view: (1) spindle cam lever, (2) motor handle, (3) motor lock lever, (4) back gear IN/ OUT selector, (5) spindle brake, (6) motor switch, (7) quill auto feed ON/ OFF, (8) quill feed handwheel, (9) quill auto feed rate selector, (10) auto feed direction plunger, (11) quill auto feed lever, (12) micrometer depth stop, (13) quill lock, (14) manual feed handle, (15) head attachment bolt.

Switching from the HI range to LO is usually no problem—no clunk from the clutch dogs. Move the spindle cam lever to raise the driven pulley, move the back gear crank to IN (or LO), then run the motor in reverse.

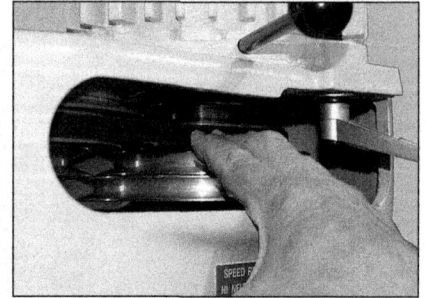

FIGURE 1-27 Clutch engagement LO to HI. Be sure the power is set to OFF; then push back on the Vee belt until you hear the clutch engage.

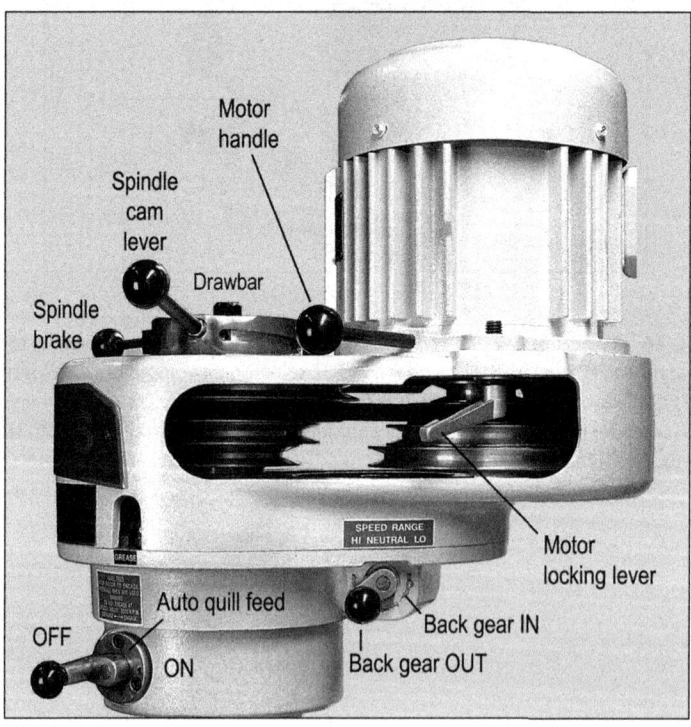

FIGURE 1-28 Typical pulley drive controls, side view.

1-19 KNEE MILL DRIVE SYSTEMS: VARIABLE SPEED

The "standard" variable-speed drive, going back to the Bridgeport days, is based on opposing cone discs that can be opened or closed by a handle at the side of the head—#3 in Figure 1-29 is typical. How this works is shown in Figure 1-30. There are a couple of caveats: Wait until the spindle has *stopped* before changing range (back gear IN or OUT), and adjust the speed only when the spindle is *running*. Finally, having changed range from HI to LO or vice versa, move the spindle back and forth by hand to be sure of proper engagement before running (that's because there are no belts to push on as in Figure 1-27).

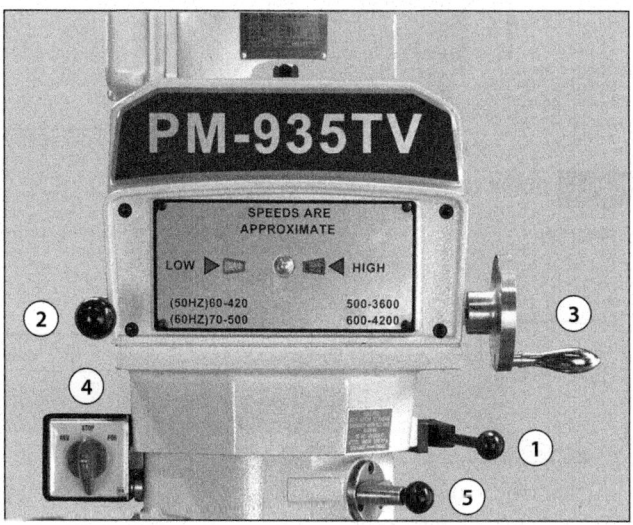

FIGURE 1-29 Typical variable-speed drive: (1) back gear IN/ OUT selector, (2) spindle brake, (3) spindle speed control, (4) motor switch, (5) quill auto feed ON/OFF.

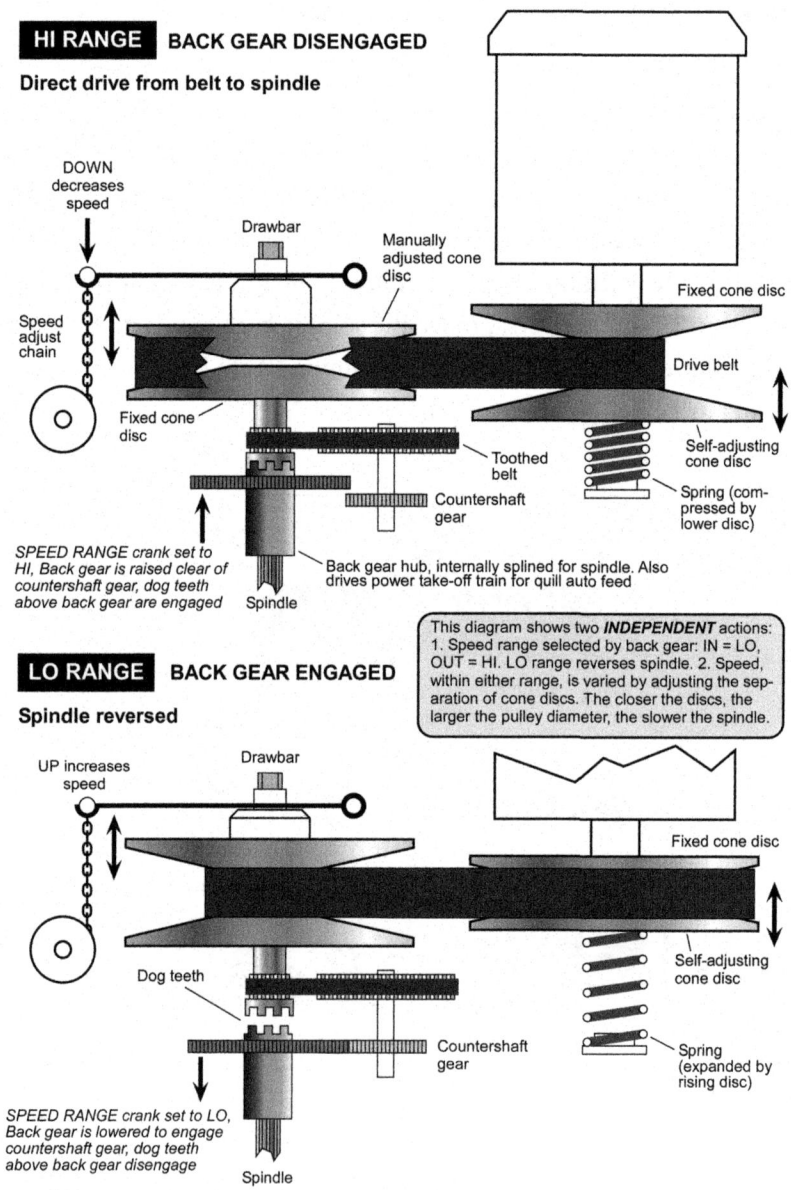

FIGURE 1-30 Typical knee mill variable-speed drive. Note the remarks in the box! The speed range selection (back gear IN, back gear OUT) is independent of the speed control (by closing or separating cone discs).

1-20 AUTO QUILL FEED

Most, if not all, Bridgeport-style knee mills come with an auto quill feed system, closely based on the years-old original design. For many users, this is an added complication they may never use, but for others it's a great time-saver in small-batch production.

It is described here for the simple reason that I have never come across anything in print about it. Aside from one or two sketchy videos on the subject, all that is known about auto feed seems to have been passed along from one machinist to the next, like folklore. The observations here are by no means definitive, but at least they are based on experience with a number of dismantled Bridgeport mill heads.

Figure 1-31 shows a typical arrangement of quill feed components. The fine-feed handwheel can instantly be removed for better access to the "direction plunger," shown separately in Figure 1-32. As its name suggests, the plunger selects UP or DOWN auto feed, plus a neutral position in the middle that disables the feed altogether. Moving the plunger from fully in

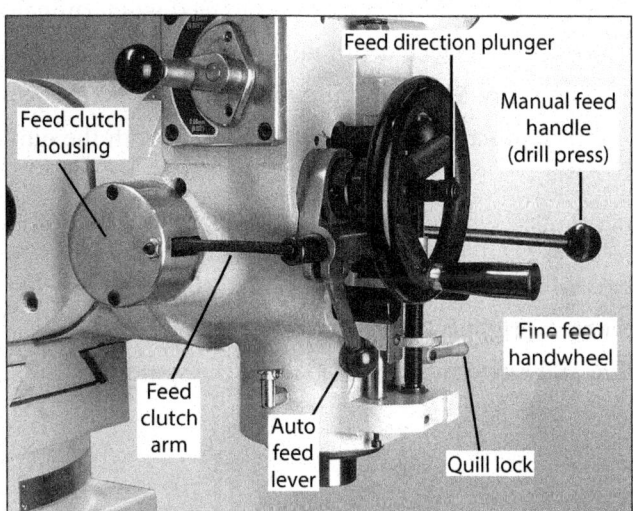

Feed direction plunger

Feed clutch housing

Manual feed handle (drill press)

Fine feed handwheel

Feed clutch arm

Auto feed lever

Quill lock

Feed rate selector. Usually 0.0015", 0.003", 0.006" per revolution of the spindle (some mills have other feed rates).

FIGURE 1-31   Quill feed components.

to fully out reverses the feed direction. Note that the feed directions are flip-flopped if the spindle is reversed, by either back gear or the motor switch.

To use the auto feed for the first time, switch off the motor, and allow the spindle to stop. Unlock the quill; then move the "drill press" handle up and down to be sure an inch or two of quill motion is available (if not, adjust the micrometer feed depth knob, shown in Figure 1-33). Select a feed rate (0.0015" per revolution for first experiments); then engage the auto feed, #7 in Figure 1-26. Start the feed action by pulling the auto feed lever out to the

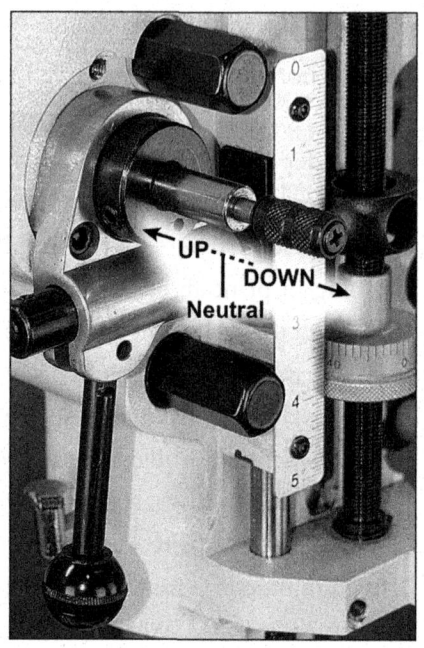

FIGURE 1-32   Auto feed direction plunger.

left. Assuming the quill motion is downward, the auto feed will terminate when the stop collar arrives at the feed depth nut, at which point the auto feed lever will trip to the right, and the quill will rapidly retract (if you prefer a less speedy return, put a *little* pressure on the quill with the quill lock lever).

Other important points:

1. The auto feed can be stopped at any time simply by pushing the feed lever to the right (vice versa to restart the feed).
2. Feed direction and rate can be changed while the spindle is running.

3. Take care not to bend or otherwise damage the (surprisingly fragile) direction plunger.

4. The auto tripping point is accurate only to about ± 0.01" (could be a lot more). For a more precise end point, hand-feed to a stop.

5. The feed system also trips off when feeding upward; see the reverse feed trip mechanism in Figure 1-34.

6. The quill can be moved using the fine-feed handwheel, even while the auto feed system is active. To do this, move the direction plunger to neutral; then place the handwheel on its axle. When the auto feed lever is pushed to the left, the hand-wheel has control of the quill.

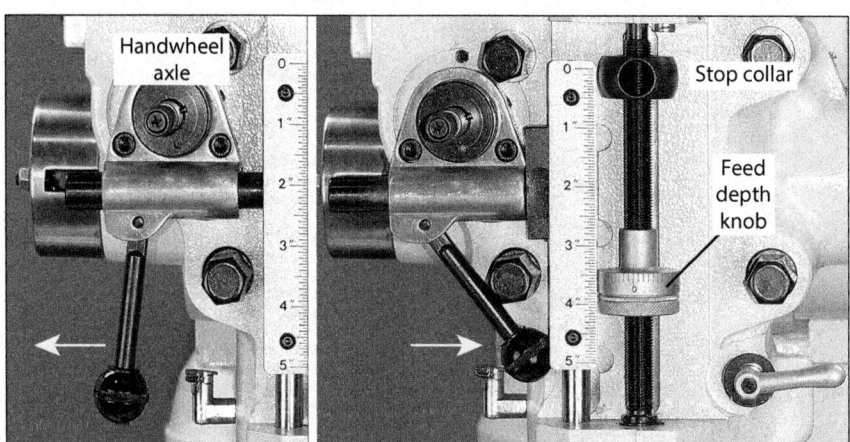

FIGURE 1-33 Activating the auto feed.

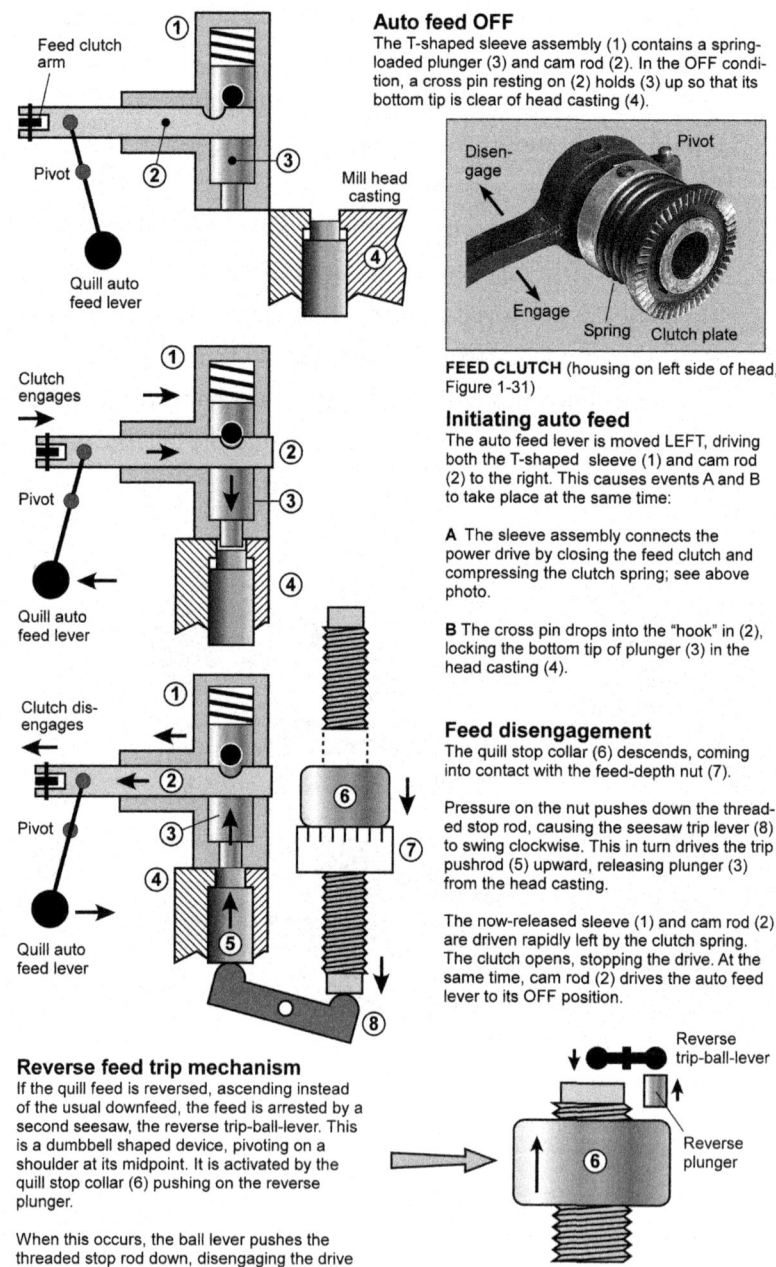

Auto feed OFF

The T-shaped sleeve assembly (1) contains a spring-loaded plunger (3) and cam rod (2). In the OFF condition, a cross pin resting on (2) holds (3) up so that its bottom tip is clear of head casting (4).

FEED CLUTCH (housing on left side of head, Figure 1-31)

Initiating auto feed

The auto feed lever is moved LEFT, driving both the T-shaped sleeve (1) and cam rod (2) to the right. This causes events A and B to take place at the same time:

A The sleeve assembly connects the power drive by closing the feed clutch and compressing the clutch spring; see above photo.

B The cross pin drops into the "hook" in (2), locking the bottom tip of plunger (3) in the head casting (4).

Feed disengagement

The quill stop collar (6) descends, coming into contact with the feed-depth nut (7).

Pressure on the nut pushes down the threaded stop rod, causing the seesaw trip lever (8) to swing clockwise. This in turn drives the trip pushrod (5) upward, releasing plunger (3) from the head casting.

The now-released sleeve (1) and cam rod (2) are driven rapidly left by the clutch spring. The clutch opens, stopping the drive. At the same time, cam rod (2) drives the auto feed lever to its OFF position.

Reverse feed trip mechanism

If the quill feed is reversed, ascending instead of the usual downfeed, the feed is arrested by a second seesaw, the reverse trip-ball-lever. This is a dumbbell shaped device, pivoting on a shoulder at its midpoint. It is activated by the quill stop collar (6) pushing on the reverse plunger.

When this occurs, the ball lever pushes the threaded stop rod down, disengaging the drive in the same way as described above.

FIGURE 1-34 Auto quill feed schematic.

1-21 BENCH VERSUS KNEE MILLS SUMMARIZED

The term "work envelope" is often used to classify milling machines. Roughly speaking, it means the maximum table travel in the left-to-right (X) and front-to-back (Y) axes, and the maximum separation between spindle and table (Z axis).

Even the smallest bench mill is capable of any milling task within its work envelope, provided you don't attempt overly aggressive metal removal. Looking beyond the machine's size, other factors to consider are:

- **Weight.** The heavier the machine, the more cast iron, the more rigid. This helps deliver smoother surfaces and deeper cuts without tool chatter.
- **Motor power.** Small bench mills have motors of about 1 hp. The largest bench mills go to 2 hp. Knee mill motors are typically 3 hp, single-phase or three-phase. Power is not usually a limiting factor, provided you adjust depth of cut and traversing speed (table motion) to suit the machine.
- **Spindle drive system.** Several small bench mills have continuously variable-speed drives—they transition smoothly from speed to speed simply by adjusting a potentiometer knob. In most cases, there is still a need to change the *speed range* by belt/pulley swapping. Most of the larger bench mills have single-speed motors and a 6-speed gearbox—perfectly functional but not as convenient. Knee mills come with either the standard 8-speed Bridgeport pulley system or a variable-speed belt drive with coned pulleys.
- **Variable-speed drive.** A 3-phase knee mill with pulley system can usually be converted to continuously variable drive simply by installing a VFD. Mechanically, everything else is left intact—use the back gear in the usual way to change the speed range.

- **Quill depth stop.** The smaller bench mills do not have this feature (surprisingly, since you'll find one on the humblest $100 drill press). This is fine if you are drilling through holes, but not so if you need holes of specific depth. However, even the smallest mills have two features you won't find on any drill press: (1) a DRO that reports quill position to a thousandth, and (2) a fine *downfeed handwheel* that helps control quill depth much more predictably than the handle ordinarily used for drilling. See Chapter 5 for more.

Most of the larger bench mills come with a quill depth stop, as do *all* Bridgeport-style knee mills.

1-22 WHICH MILL TO CHOOSE?

The right choice of mill is entirely personal, depending on the type of project you have in mind, shop dimensions, and—above all—budget. Based on my having used most sizes and types of model shop mills over the past 20 years, my advice is to start out with a small, low-cost machine, and then graduate to a more capable mill when there's a clear need—and you have a good grasp of what really matters. I have never had a problem in selling a used machine on a "buyer collects" basis. (But building a shipping container is a whole other matter.)

On the other hand, if your budget can handle it, and your shop floor can take the load, you cannot go wrong with a knee mill from day one—assuming you've been assured of support and spares service. As generations of machinists will attest, there is really no comparison with a Bridgeport-style machine.

No matter which product you opt for, once you have it delivered, it's basically a fixture weighing 400+ lbs., very difficult to return to the supplier. Aside from the return shipping cost, which could be $1,000 for a knee mill, a shipping container is necessary. So, until you're sure the mill is a keeper, the best way to keep return costs under control is to save the container it came in—with the minimum of nail tears, etc. In other words, don't rip it apart.

DOVETAIL GIBS ON THE TABLE SLIDES

We take for granted that a milling machine's table, its saddle, and headstock (or knee) will move smoothly, yet be firm enough to resist cutting action. In other words, they must be adjustable for the best compromise between ease of motion and firmness.

A milling machine has three dovetail male/female assemblies, one each for the table's X and Y axes, plus one for the Z axis (headstock or knee). The male dovetail is narrower than the female, leaving a gap that is filled by a *gib* strip; the gib strip (aka "gib"), usually made of cast iron, has a parallelogram-shaped cross-section to mesh 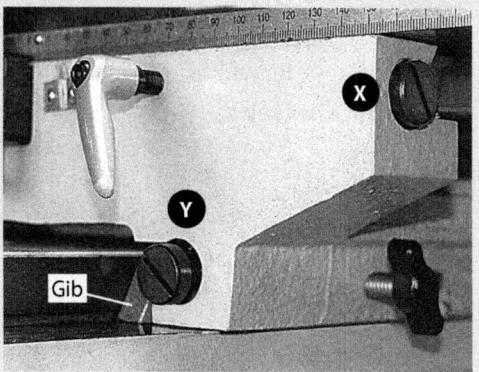 neatly with the dovetail angles. On smaller machines with conventional (parallel) dovetails, the gibs are *uniformly thick* from end to end; they are held in place by three or more set screws which are tightened for a solid, sliding fit between male and female halves.

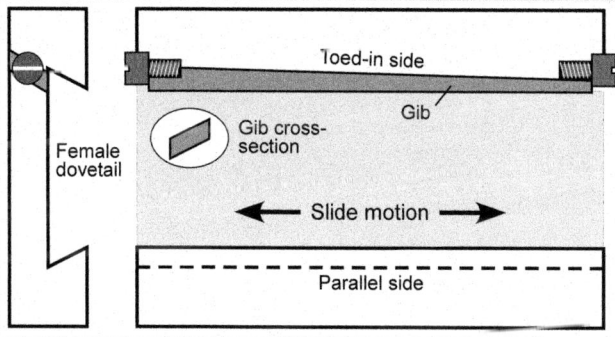

Tapered gib on large mills

On larger machines, both the dovetails and the gibs are quite different. One side of the female dovetail on each axis is machined with a "toe-in" relative to the opposing side, greatly exaggerated in the diagram here (in practice the taper is usually only a little over 1°). Like those in the smaller mills, the corresponding gibs are cast iron, and parallelogram-shaped in cross-section. But there the similarity ends—the gibs are wedge-shaped, *precisely tapered* to match the toe-in, and are held in place by two large-headed special screws, one at each end, snugged against the gib to prevent side-to-side movement. The screw heads are either cross-cut, as here, or socketed for a hex key.

CHAPTER 2

Installing a Milling Machine

CONTENTS AT A GLANCE

2-1 A VOYAGE OF DISCOVERY

Installing a mill is often exactly that, unless your machine is one of the mini variety that can be lifted by hand. If you plan to use an engine hoist, you need to be sure the entire floor area can handle the load, not just the mill's working location. The hoist needs to roll smoothly, under load. At the very least, the floor in the working location must be able to support the total weight of the mill, plus its bench if necessary, plus an additional 300 lbs. or so for accessories and the material you plan to be working on.

2-2 FLOOR STRENGTH

Poured concrete is always the best option, but not all poured basements or garage floors are thick enough—unless reinforced—to withstand a load of a ton or more. There is often no way of finding out definitively, even if the building codes are reassuring (you hope they were complied with). If in doubt, think of ways to distribute the load with *steel angle* spanning the bench feet, or expanding the footprint of a knee mill casting.

2-3 CEILING HEIGHT AND LIFTING?

This is the other key factor, which applies to all hoist-assisted installations. There may be obvious ways of dealing with this "height-of-hoist" issue, such as shortening the sling and chain, but the fact is that every install job calls for some degree of planning and ingenuity. In Figure 2-1, for instance, the mill was shipped preinstalled on its stand, so the first question was how to remove it from its pallet. It was necessary in this case to place the engine hoist legs straddling the pallet. This in turn called for part-dismantling of the hoist, which was then raised a foot or so on stacks of lumber.

In the more typical case—mill and stand shipped separately—there is often still the need for similar workarounds to remove the mill from its pallet and also to raise it to clear the stand, as shown in Figures 2-2 and 2-3. In all these hoist illustrations, the webbing sling was installed "basket style" under the headstock, with padding and wedges where necessary to avoid damaging the paintwork and/or fragile components, shown in Figure 2-4.

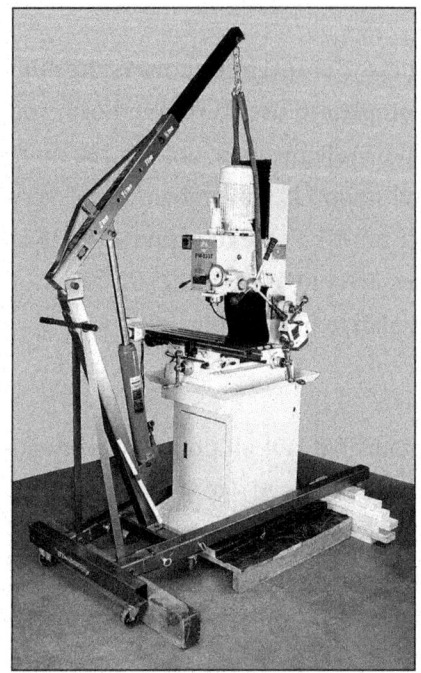

FIGURE 2-1   Engine hoist issues.

FIGURE 2-2   Lowering onto the stand.

On some of the larger bench mills, eyebolts are installed on the sides of the base casting (Figure 2-5). These are intended to be used with steel lifting rods (about 3' long x 1"-diameter steel) and a forklift.

Balance the mill before lifting! *Lower the mill's center of gravity by lowering the headstock. For balance, move the table back toward the column. Lock all movable components.*

2-4 SITING AND LEVELING THE MILL

The most obvious consideration, mostly for bench mills, is clearance for the moving components—table left to right and front to back, headstock fully elevated. Less obvious is the need for access to all parts of the mill for maintenance, such as the back and top of the column.

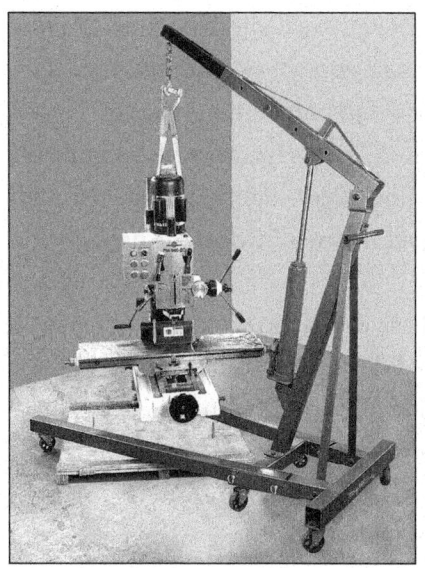

FIGURE 2-3 Removing a bench mill from its pallet.

FIGURE 2-4 Keep the sling clear of all fragile components, in this case the electrical box.

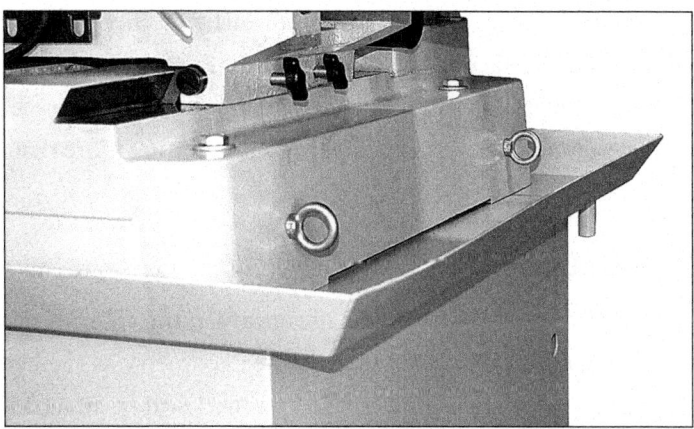

FIGURE 2-5 Lifting eyes.

All bench mills need to be securely bolted to a dedicated stand or bench. A bench, if heavy and rigid, may be stable enough without floor anchors. This may also apply to a dedicated stand, but only if it is sufficiently heavy (cast iron), not a sheet metal construction, unless heavy gauge. Bolting the stand to the floor is also highly recommended—lightweight bench mills tend to be unstable when the table is out to one side or if the headstock is fully raised.

Bear in mind that a bench mill's base casting may not be as rigid as it looks, and it may be distorted if bolted down to the stand or bench. The remedy for this is to shim the underside of the base casting before bolting down. The same applies to the underside of the stand—shim under the stand before bolting down on the concrete.

Except in commercial shops, where local ordinances may dictate otherwise, knee mills can usually be operated safely without floor anchors. In every case, including bench mills, leveling the table in both axes is always worthwhile. While it is true that machining accuracy isn't greatly affected by leveling (as it can be with a lathe), a level table can be a useful point of reference when setting up the workpiece. For instance, if you need to skim-cut a rough-sawed surface with the minimum of waste, set the workpiece in the vise; then tap its top surface up or down for the "best average" across the saw-cut peaks and valleys, gauging this with nothing more than a standard carpenter's level.

2-5 BENCH MILL RIGIDITY

In bench mills, it is possible that lack of rigidity in the column, and other factors, may cause variations in the machine's geometry as the headstock is raised and lowered. (Imagine a laser pointer, firmly attached to the headstock, striking a specific spot on the table; in theory, but only in theory, the laser beam will not deviate from that spot if you press against the headstock, even at the higher headstock elevations.) This is something to be aware of, but it is not likely to be a serious issue in model shops where the workpiece and spindle are within a few inches of each other. Much *more important* in every case is that the spindle should be perfectly at right angles to the table when the headstock elevation is within the normal working range. One word for this is "tramming," which is a routine shop procedure (see Chapter 4, Section 4-2).

2-6 CONSIDER PARTIAL DISASSEMBLY

This applies mainly to bench mills. If a hoist is not available, and the mill has to be moved by manpower, it may be more manageable if the headstock and table are temporarily removed. Removing the table is straightforward, usually a matter of removing components from both ends of the lead screw, followed by the gib from the front dovetail. The table can then be slid to one side and lifted clear.

Removing the headstock, at 200+ lbs., may be a two-person job. Makes and models of bench mills vary considerably, but in most instances the headstock is secured by three bolts in a 120° pattern. In Figure 2-6, the procedure was as follows:

1. Assemble a stack of 2 x 4s.
2. Crank the headstock down to the point where it is felt to be pressing firmly on the back 2 x 4s—*firmly*, meaning the headstock will not suddenly shift when its attachment nuts are loosened.
3. Lower and lock the quill so that the spindle nose rests on the front 2 x 4 (this is for balance, not for load bearing).
4. With another person on hand to stabilize the headstock, remove the attachment nuts; then move the table forward clear of the T-bolts (which usually remain in the sliding casting).
5. If necessary, use a telescoping magnetic pickup to retrieve/reposition T-bolts, as shown in the inset in Figure 2-6.

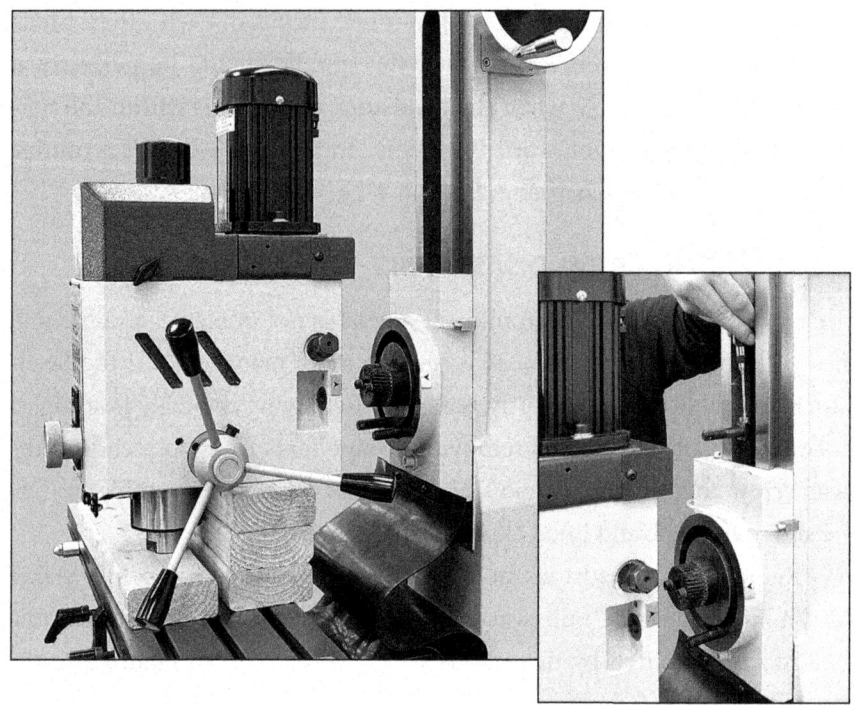

FIGURE 2-6   Removing the headstock. The inset at the right shows a telescoping magnetic wand being used to capture a T-bolt.

2-7 HANDLING A KNEE MILL

Aside from their much greater weight, knee mills are generally more predictable in handling. For instance, because they are floor-standing, they need only be raised a couple of inches before being rolled into position.

Most knee mills have provision for an eyebolt on the ram (Figure 2-7). In commercial shops, this is usually chained to the tines of a forklift (Figure 2-8). It can also be used with an engine hoist, but take care not to exceed the hoist's lifting capacity. If there is no eyebolt, run a sling (basket style) under the ram, as shown in Figure 2-10. Both methods call for careful balancing, and packing materials between mill and forklift.

If there are clearance or balance issues when moving the mill to its working location, it may be necessary to swivel the head and/or tilt it for-

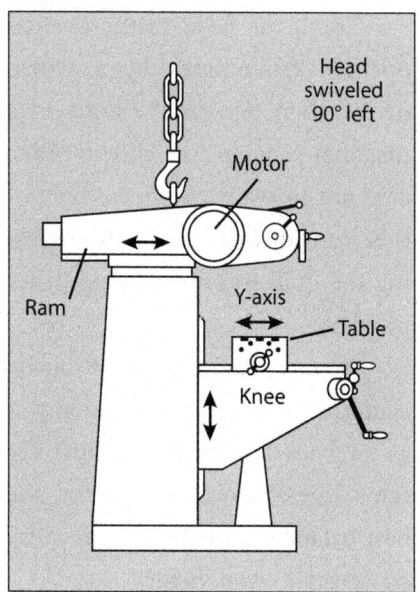

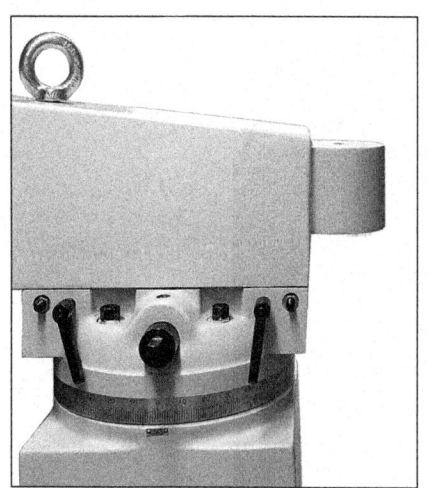

FIGURE 2-7 Knee mill ram with eyebolt.

FIGURE 2-8 Hoisting by eyebolt. The head is shown here swiveled 90° counterclockwise.

ward (Figures 2-8 and 2-10). If swiveling is necessary, partially loosen the four head mounting bolts (Figure 2-9).

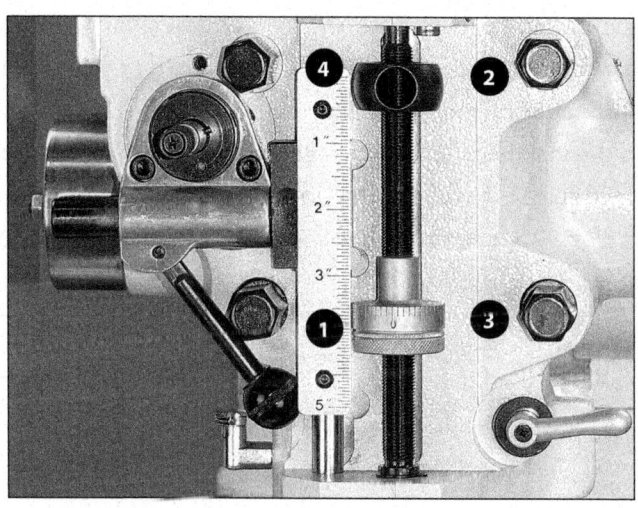

FIGURE 2-9 Head mounting bolts.

Crank the head to the desired position (sometimes by a worm drive); then tighten the bolts in a diagonal sequence as shown. Take care not to overtighten the bolts—this can distort the head, causing the quill to bind, among other problems.

Prepare the mill for lifting by centering the table and moving it as far back as possible. Adjust the ram's fore-aft location to find the best balance point for the lift. *Lock all movable components.*

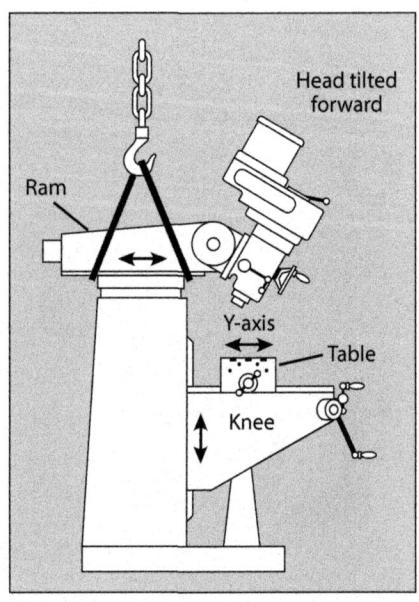

FIGURE 2-10 Hoisting by slings. Move the ram in or out as necessary. Protect the ram ways with padding. Tilt the head forward for clearance if needed.

2-8 DO IT YOURSELF, OR GET HELP?

Sometimes it makes sense to call in professional help when the job is too much to handle, even with the help of a neighbor or two. In most parts of the country, there are machine installers ("riggers") who will do the job as a matter of routine. It may be possible to have the machine(s) shipped to the installer's address, as opposed to having a massive crate (or two) dumped in the driveway. The downside is cost, which will probably be upward of $500. Personally, I have opted for professional help on several occasions and would do it again.

RUBBER MATS CAN AVOID DISASTERS

Rubber mats around benches and machine tools can save the day when you accidentally drop something fragile, delicate, or sharp—think dial indicators, calipers, milling cutters (the mats also provide foot relief). Look for 3/8"- to 1/2"-thick material, ideally nonslip.

CHAPTER 3

Things to Know from the Start

CONTENTS AT A GLANCE

3-1 WHAT THIS BOOK DOESN'T TELL YOU

Because the focus of this book is on machining, it's assumed you don't need to hear the basics of manual shopwork such as layout, cutting to size, drilling, filing, and so on. The same goes for height gauges, surface plates, gauge blocks, etc. That said, there are a few areas where "shop detail" is included, if only because it can be very relevant to to milling operations and may not be fully explained in other shop publications. Examples of these areas are scales and micrometers; hole diameter and depth measurement; filing and edge deburring; and common steel and stainless alloys.

Information specific to your model of mill, such as lubrication and maintenance, should come with the machine. If no such docs were provided, bear in mind that all machine ways (usually dovetails) need careful oiling with a "tackified" oil such as Vactra No. 2. Use a flux brush to apply (shown in Figure 3-21 in Section 3-22). Most rotating parts run in pre-greased ball or roller bearings and should need no routine attention (but you need to ask your supplier).

3-2 WHAT MILLING IS ALL ABOUT

Milling is basically the process of removing material from a block, a sheet, or a bar of metal. It helps if the workpiece starts out in rectangular form, with at least its reference faces nicely squared with respect to each other. That's certainly how it could be if you are shopping for a specific project and have the metal supplier cut off an inch or two of this or that bar stock. Most of us, though, start with whatever material is on hand, in some cases tweaking designs to avoid going out for additional supplies. Inevitably, this calls for additional prep work, first cutting the stock to size on the bandsaw, then machining it to a suitable shape from which to proceed.

3-3 THERE'S NO SUCH THING AS A SIMPLE JOB

There is a process—a specific order of events—to even the most basic milling job, such as squaring off an exact 2" length of rectangular bar stock, say 1" x 1/4" aluminum. One way to tackle this is the following:

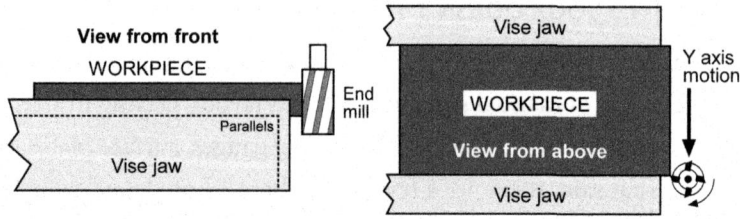

FIGURE 3-1 Workpiece motion versus cutter rotation.

1. Check and adjust the vise for squareness relative to the table.
2. Cut the stock overlong by about 0.1" using a hacksaw or bandsaw; then clamp the material in the mill vise, raising it on parallels.
3. Install a 3/8" or 1/2" end mill; then lower and lock the spindle.
4. Adjust the table to bring the cutting edges into contact with the right edge of the workpiece. Why the right edge? Because in the cutting pass, the workpiece should be moving forward (toward you), *opposing* the clockwise direction of the cutter teeth. (The other direction is called "climb milling," not recommended for first-time experiments.)*
5. Lock the X axis; then bring the workpiece forward (Y axis), to skim off a few thousandths of an inch of material, and then reverse Y to the starting point.
6. Advance the X axis a few thousandths, relock, then make another pass. Repeat as necessary for a fully machined surface
7. Remove the workpiece; then deburr the machined end (because raised edges and chips cause inaccurate measurements).
8. Flip the workpiece in the vise; then skim the other (rough-sawed) end in the same way. After making the close-to-final pass, zero the X handwheel dial *without moving the handwheel*.

---

*If instead the end mill is on the left side of the workpiece, the cutting pass starts at the back edge, and the table moves backward (away from you). The same thinking applies to the X axis, cutting front versus back edges of the workpiece.

9. With the workpiece still in the vise, measure its length using calipers. If the length is 2.035", another 35 thousandths of an inch has to be removed.

10. Advance the X axis by 30 divisions on the dial; then make another cutting pass.

11. With the workpiece in place, deburr the just-cut end; then recheck the length.

12. Make a further very small adjustment to the X axis; then skim again to bring the length to—we hope—exactly 2" (more realistically, we should be aiming for 2 ± 0.001").

Lengthy though all this seems for a rather basic job, you will find that these actions soon become almost instinctive and obvious. For any job outside the obvious range, though, the *golden rule* is to check all measurements and calculations, then recheck them before cutting metal (*measure twice, cut once*). Don't just glance at the drawing, even if it's a sketch you've just done. Take another careful look. Try to visualize the entire process; don't paint yourself into a corner.

Don't start the cut until you're sure everything's tight and solid—vise T-nuts, spindle drawbar, and table and headstock locks. Make sure the workpiece is firmly gripped by the vise (or clamped to the table). Finally, don't loosen the workpiece until you're sure that specific portion of the job is done.

3-4 SCALES AND MICROMETERS

In the woodshop and around the home, you probably use scales graduated in fractions of an inch: 1/4", 1/8", 1/16", etc. That's because we are very familiar with such fractions, and the instructions we're working to are usually dimensioned that way. The same scales are not as useful in the metal shop, where measurements are more likely to be in thousandths of an inch (mils) or millimeters. This calls for different measuring tools, starting with the "engineer's scale." The most useful choice is generally a *flexible 1/2"*-

wide scale, 6" or 12" long with decimal/metric graduations: 1/10" and 1/50" on one side, 1 mm and 1/2 mm on the other (Figure 3-2). Flexible scales are between 0.015" and 0.02" thick, which gets you closer to the workpiece than the 3x–thicker rigid scale. Decimal/metric scales like this are worth shopping for—usually online only.

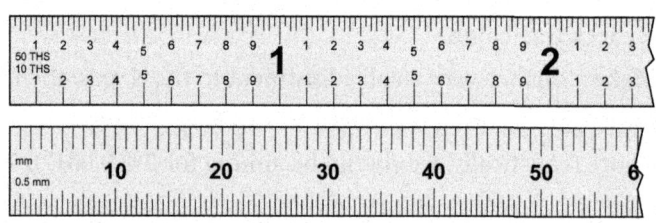

FIGURE 3-2 Typical inch/metric engineer's scale, front and back.

With a "fiftieths-of-an-inch" scale like that shown in Figure 3-2, you can directly measure to the nearest 0.02". With interpolation, you can estimate to ± 0.01", 10 thousandths, but that's the practical limit. On the metric scale, we are limited to the resolution of the finest graduations, namely 0.5 mm. To do better than that, we need a scale with a more precise *interpolator*, in other words, a vernier scale. This was invented almost 400 years ago by a French mathematician, Pierre Vernier. It has been in global use ever since.

How a vernier scale works is an interesting mystery to many of us, including experienced machinists. The following paragraphs may help.

First, imagine a generic "decimal-reading scale" caliper graduated in any unit you wish—feet, inches, or centimeters.

Look at Figure 3-3. Note that the jaws are closed, with both zeros aligned. And note that 10 on the vernier scale spans 9 on the main scale, a one-count difference that is typical of all vernier scales.

Now separate the jaws so that the vernier 0 is about three-quarters of the way between 0 and 1 on the main scale (Figure 3-4).

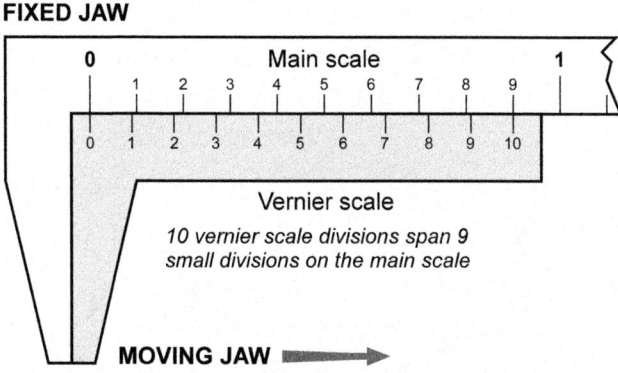

FIGURE 3-3 Generic vernier caliper, fully closed.

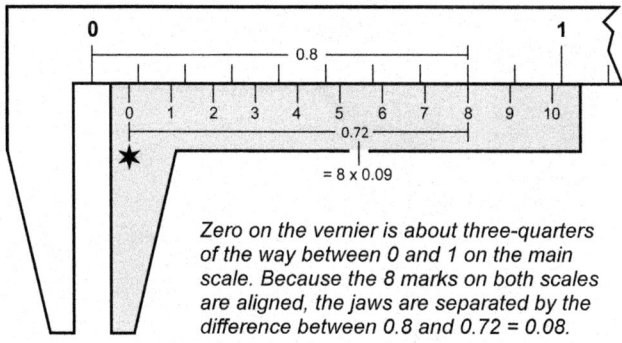

FIGURE 3-4 Generic caliper, jaws separated.

How can we know more exactly where the vernier 0 is, which tells us how far the jaws are apart? The answer is to look along the vernier scale for a line that coincides with a line on the main scale. In Figure 3-4 it's line 8, so the jaws are separated by precisely 0.08 on the main scale. In practice, the scale would be graduated in decimal fractions of an inch or in centimeters, so the 8 could be read as a specific measure of distance.

Thinking of "measure of distance," there are questions that always crop up when you use an unfamiliar vernier scale: *How can I interpret what I'm seeing here? What is the scale's resolution?* Sometimes this is announced (for instance 1/1000", upper scale in Figure 3-5), sometimes not. If not, determine the unit of measure (U) represented by the smallest division on the main scale, usually 0.025" or 0.05" on an inch-reading caliper. Then count the number of divisions (V) on the vernier scale. Finally, divide U by V.

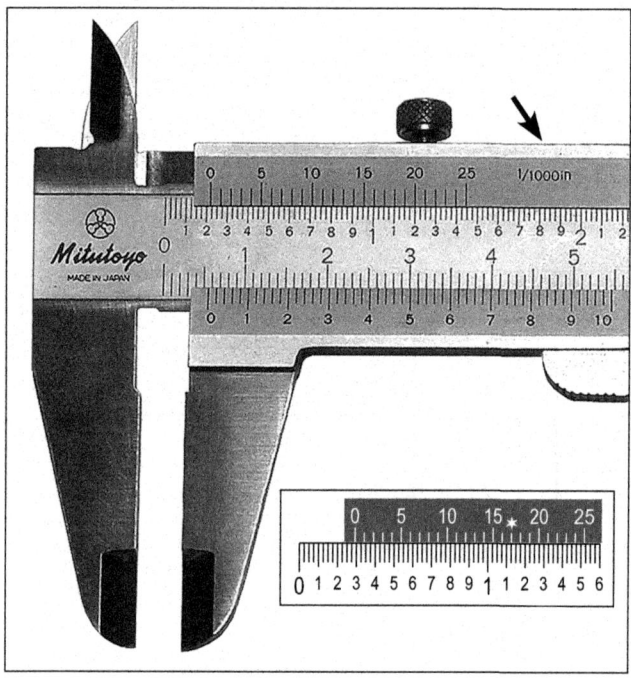

FIGURE 3-5   Inch/metric vernier caliper. The scale resolution is sometimes stated, as indicated by the arrow.

In the real-life vernier of Figure 3-5, the upper scale is in inches, divided into tenths, with further subdivisions (U) of 0.025". The vernier scale has 25 divisions (V), so the resolution, U/V, is 0.025"/25 = 0.001" (1/1000, as etched on the scale at top right). In the figure, the vernier 0 is just to the right of 2 on the upper main scale, meaning a little greater than 0.2". Further to the right, there is perfect coincidence of the two scales at vernier 20, so the jaws are separated by 0.2" + (20 x 0.001") = 0.220".

In the second example, Figure 3-5 inset, the vernier 0 is just to the left of 0.3" on the main scale, so you can see that the measurement is the total of three sub-sub-divisions, plus the amount indicated by the vernier. There is coincidence at vernier 17, so the total measurement is 0.2" + (3 x 0.025" + (17 x 0.001") = 0.292".

The lower scale, metric, is quite different. For one thing its resolution is not announced, so we need to figure it for ourselves. Here 50 divisions (V) on the vernier span 49 divisions on the main scale—the usual one-count difference. Each division of the main scale represents 1 mm (U), so the resolution, U/V, is 1 mm/50 = 0.02 mm (a shade finer than 0.001"). In Figure 3-5, the vernier 0 is about midway between 5 and 6 mm on the main scale, meaning that it seems to be in the range of, say, 5.5 through 5.7 mm. Looking to the right, there is perfect coincidence at vernier 6, so the correct value is 5.6 mm. This brings us to the conclusion that numbers on this particular vernier *directly report* tenths of a millimeter (with subdivisions of 0.02 mm).

The takeaway here is that Step 1 in all vernier measurements is "visual interpolation" to determine roughly where the vernier 0 lies, followed by Step 2, careful checking for coincident lines.

Both steps call for a steady hand, good lighting, and good eyesight, all factors that spawned easier-to-use technologies—first the dial caliper, then the direct-reading digital scale.

So why trouble with a centuries-old measurement method? Three answers: First, because there are great deals out there for vernier calipers and height gauges. Second, there is nothing to go wrong with a vernier caliper—it's coolant-proof and has no battery and no fragile, easily jammed rack and pinion. Third, as will be described later, because many *standard inch micrometers* use a vernier scale to increase resolution from thousandths of an inch to 10 thousandths, 0.001" to 0.0001".

Verniers are also used for angular measurement in various lab instruments, and in *precision bevel protractors* often used in the machine shop. You will probably have realized already that there is nothing obvious about reading any vernier, much less so in the case of angle verniers. Because how-to explanations are hard to come by, we take a brief look here at angular measurement before returning to linear measurement.

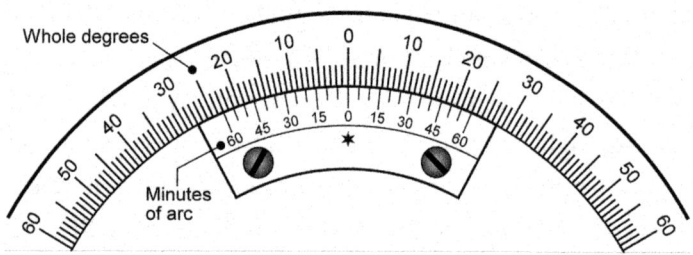

FIGURE 3-6   Bevel protractor. Main scale and vernier zeros aligned.

The main scale of this protractor is graduated in degrees, Figure 3-6. Below the main scale is the typical "double" vernier scale, symmetrical on both sides of the vernier zero. On each side of the centerline there are 12 vernier divisions, each spanning 23 degrees on the main scale. Each vernier division therefore spans 23/12 degrees; this equals 115 minutes of arc, which is 5' less than 2 whole degrees (120') on the main scale. This means that the resolution of the protractor is also 5', or 1/12 of a degree (compare this with the U/V calculation described earlier for linear verniers).

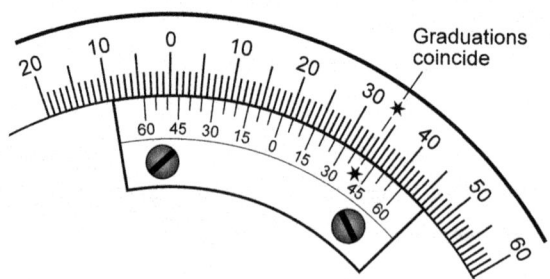

FIGURE 3-7   Vernier scale rotated to the right.

In Figure 3-7, the vernier has been rotated *clockwise* relative to the main scale, bringing the vernier zero to just over one-half the distance between 18° and 19°. In the same *clockwise direction* (important) the only vernier mark that aligns with the main scale is 40, so the angle measured is 18°40'.

Now, back to linear measurement, moving on from the scale caliper to the micrometer—less flexible, but often thought to be the more reliable.

In the inch-reading micrometer in Figure 3-8, the numerals 0, 1, etc., above the left-to-right datum line represent intervals of 0.1" with four subdivisions of 0.025" (these are not numbered—your first hint that the micrometer calls for visual interpolation, just like the vernier caliper). The rotating sleeve is numbered **0** 1 2 3 4 **5** 6 7 8 9 **10** 11 12 13 14 **15** 16 17 18 19 **20** 21 22 23 24 **0**, each division representing an interval of 0.001". In the figure, the total distance visible along the datum line is 5 x 0.025" = 0.125".

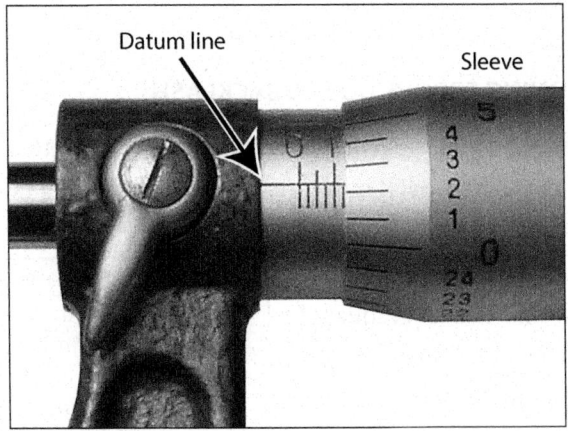

FIGURE 3-8 Inch-reading micrometer.

If the 0 mark on the sleeve happened to coincide with the datum line, the measurement would be exactly 0.125". However, the line is between the 2- and 3-mil lines on the sleeve, in fact just a shade beyond 2. If we are working only to the nearest thousandth, we would say that the measurement is 0.125" + 0.002" = 0.127". If we need to be more precise, we lock the sleeve, then roll the micrometer forward to view the vernier (Figure 3-9). Here, the only vernier line coincident with any line on the sleeve is 3, indicating that the measurement is 0.1273".

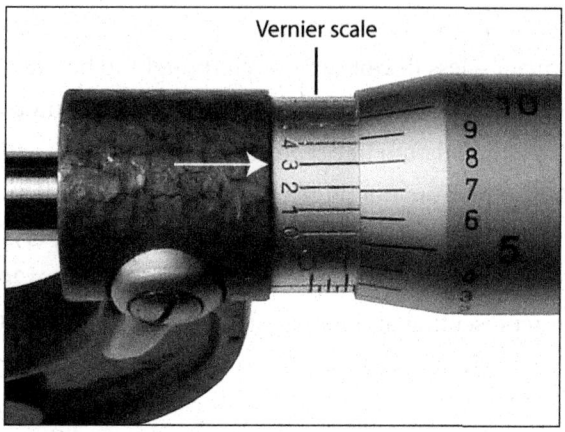

FIGURE 3-9   Micrometer vernier scale.

3-5 POSITIONING ERRORS DUE TO BACKLASH

Longitudinal, left-right movement of the table is defined to be the X axis. Crosswise travel, front-back movement, is the Y axis. The table is moved by two accurately machined lead screws, usually 10 TPI (threads per inch). Each axis has a graduated collar with 0.001" divisions, 0.1" per revolution (Figure 3-10).

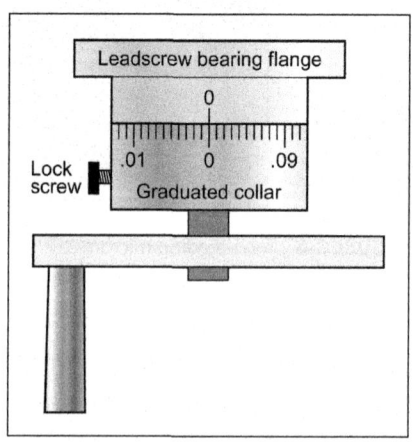

The table can be positioned precisely by counting full turns of the lead-screw handwheel, plus a part turn measured by divisions on the collar.

Example: A movement of 0.519" calls for five turns plus 19 divisions.

FIGURE 3-10   Lead-screw handwheel and collar.

Watch out for *lead-screw backlash*, which is due to loose coupling between the lead screw and table. In some cases, it can mean that reversing the lead-screw handwheel by as much as 20° or more has *no effect whatever* on table position.

To avoid the effect of backlash, table motion must always be in the *same direction* when moving from one location to another. Suppose you need to drill a row of holes—A, B, and C—at 1/2" intervals. You would drill A and make five turns of the handwheel; drill B and make five more turns; and then drill C. You cannot go from A to C, then back to B, expecting the same 1/2" separation: B to C will be at least 0.005" short. This is not an issue if a DRO (digital readout) is installed.

3-6 POSITIONING THE TABLE BY COUNTING DIAL DIVISIONS

Aside from lead-screw backlash, discussed in the previous section, there is one other factor to keep in mind—errors caused by *locking the quill*. On most vertical mills, including knee mills, locking the quill may offset the spindle by 1 or 2 thousandths of an inch. This means that all position-finding operations should ideally be done without locking. If you can, retract the quill fully; then lower the quill with the fine downfeed control. On bench mills, this is worm-driven, so it stays where it's put without locking. If working on a knee mill, raise the knee instead—again without locking the quill. *However*, for operations needing depth control, such as milling, the quill should be locked to maintain the cut depth.

The following "Y axis position-ing demo" assumes a table motion of 0.1" per turn of the handwheels—yours may be different.

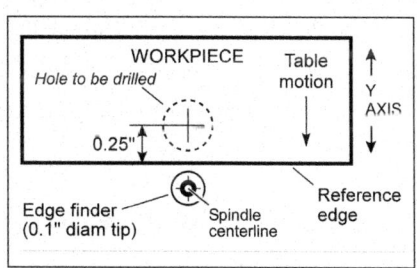

As illustrated in Figure 3-11, a hole is to be drilled 0.25" on the Y axis, relative to the front edge of a workpiece held in the vise or other-wise clamped to the table.

FIGURE 3-11 Spindle positioning example.

Here are the steps:

1. Install an edge finder in a collet or chuck (a tip diameter of 0.2" is assumed).
2. Lock the X axis by tightening both levers.
3. If the reference edge is already to the back of the spindle center-line, do nothing; if not, rotate the Y axis handwheel (usually clockwise) to send the workpiece backward, until the reference edge is behind the spindle centerline, as shown in Figure 3-11.
4. Engage the fine downfeed.
5. With the spindle running at 500+ rpm, lower the quill as necessary using the fine downfeed handwheel; bring the table forward, stopping at the point where the edge finder just makes contact (the tip jumps out of line). Raise the quill, and/or stop the spindle.
6. While holding the Y axis handwheel to prevent movement, zero the dial; then retighten the dial thumbscrew.
7. Raise the quill; then rotate the handwheel one exact full turn counterclockwise (0.1") to bring the reference edge forward to coincide with the spindle centerline.
8. Rotate the handwheel 2-1/2 additional turns counterclock-wise to bring 50 on the dial opposite the datum; the spindle is now exactly 0.25" behind the reference edge.

Time-consuming though it is, "positioning by counting divisions" is what machinists everywhere did until just a few years ago—and many of them still do. The game-changing enhancement for the fortunate among us is the DRO, which eliminates the backlash issue completely (see Chapter 7).

3-7 WHERE TO FIND ACCESSORIES AND MATERIALS

This is a question that comes up in practically every conversation having to do with the model shop. There are not as many machine tool suppliers as there were 20 years ago, but those that remain are very active.

Four examples:

1. Precision Matthews PA (lathes from 10" x 22" to 14" x 40", bench and knee mills, table widths from 27" to 54")
2. Little Machine Shop, CA (small lathes and mills, many accessories)
3. Wholesale Tool, MI (formerly Victor Machinery, NY—wide range of accessories, measuring equipment and cutting tools)
4. Grizzly Industrial, WA and MO (wide range of wood and metalworking machines and accessories).

For peace of mind, you need to be comfortable with your supplier's product support—*will you be able to talk to an actual user of the machine?*

For materials there is usually no option other than buying online, because most local metals stockists have disappeared. Online suppliers offer a good range of metals in the usual choices of alloys, shapes, and sizes. The downside of online buying is the cost of freight, which can be startling.

One big-name supplier you can rely on to stock just about everything to do with machines is MSC. Another reliable supplier is McMaster Carr, for many years the go-to source for every conceivable item of hardware, including metric. You can also go to McMaster Carr for a very wide range of materials, available overnight if you need it. *One caveat:* The company does not (or used not to) quote freight costs.

3-8 CHOOSING THE RIGHT KIND OF STEEL

There are two main classes of steel—hot rolled and cold formed (or cold rolled). Hot rolled is the product found in steel construction shops where the material is typically flame-cut and finished—if at all—by grinding. The most common hot-rolled steel alloy is A36. It is machinable—sometimes not as easily and cleanly as the most common cold-formed alloy 1018, which is similar in composition to A36 but with slightly lower carbon content. A36 is generally less expensive than 1018, maybe because of the higher volume of production and distribution.

1018 steel is available in some hardware stores in sheets from 1/16" thick and up, and also in square, rectangular, and round sections up to 1/2". For much better choices of thickness and section, go to a metals stockist. There may be one within driving distance, but if not, any steel you need can easily be purchased online—the downside of that, as noted earlier, is shipping cost.

Before buying anything, consider a free-machining alloy such as 12L14 (trade name Ledloy) or 1215 as a lead-free alternative. The difference in machinability and surface finish compared with 1018, and practically every other alloy, is surprising. It can be used in most applications as a direct substitute for 1018. Additionally, 12L14 and 1215 can be case-hardened for some degree of wear resistance.

The only downside with free-machining steel alloys is shapes and sizes: They are usually available only in bar form (not sheets) in square, round, and hexagon sections. Thin sheets and rectangular sections can occasionally be found, but don't bank on it.

Table 3-1, in Section 3-23 at the end of this chapter, lists selected steel and stainless-steel alloys.

3-9 A LOW-COST MODELING METHOD

A trial run can avoid mistakes! Like everything else in engineering, the finished product is the best compromise of competing factors.

A reliable way to be sure you have the dimensions right is to model the entire project in materials that are easy to work—cardboard, wood, or (much better) an "engineering material" like rigid PVC. This is a low-cost plastic, available in rod, bar, and sheet from many suppliers, including McMaster Carr.

PVC machines beautifully at high spindle speeds (Figure 3-12), with no significant cutter wear—but watch out for overheating. Blanks of the material can be rough-cut to size in seconds using a table saw with a wood-cutting blade. Aside from solid-modeling CAD and 3D printing, this is the fastest way to a realistic fit, form, and function model (with a lot less learning time, too).

FIGURE 3-12 Machining PVC is quick and easy.

Another bonus you get from working in PVC is the ability to practice machining procedures with the least possible risk. *One caveat:* When heated, PVC expands four times more than steel—and it cools more slowly—so take care when checking measurements. Reliable screw cutting can also be an issue, especially with fine-pitch threads.

For experimental work, I keep a stash of 3/4"-thick dark-gray PVC (Type 1), plus a few oddments of bar stock. Where extra thickness or composite shapes are needed, PVC can be bonded in a minute or two using plumbing adhesive.

3-10 DO I NEED CUTTING OIL?

In *most* cases, *no,* if you are machining plastics or taking very light cuts on most metals, including general-purpose low-carbon steel (less than 0.2% of carbon). This applies also to cast iron (about 4% carbon), which machines quite differently from other metals—expect powder instead of chips at the cutting tool. Things change as you become more adventurous, taking deeper cuts to save time. Still no need for cutting oil on brass and aluminum, but it will give you better results on steel. On steel, one thing it will definitely do is save wear on the cutting edge. It may also improve surface finish.

For first experiments try one of the off-the-shelf cutting oils that come in little bottles (Tapmagic, Tapfree, Cool Tool, Relton, etc.). Squirt a little oil into a disposable plastic cup and apply using a brush of the sort used for solder flux (also disposable, but in this application, it can last forever).

3-11 INDUSTRIAL CUTTING FLUIDS

In some instances, you will get better results with the cutting fluid used in commercial machining, on practically every metal, including aluminum. This is a *water-miscible* cutting fluid such as Blaser Swisslube. You might be horrified to think of spraying a thin white fluid that's mostly water on your favorite machine, but machine shops do that all day, every day, on their CNC (computerized numerical control) machines.

Water-based cutting fluids perform two important functions:

1. They assist cutting action, just like Tapmagic, etc., often helpful when parting off and performing other troublesome operations.

2. They cool the cutting tool and workpiece. You won't hear much about water-based fluids in model engineering books because the concentrate is sold only in expensive multi-gallon containers. Another factor to be aware of is the mix ratio, which has to be precise—something on the order of one part concentrate to ten parts water. If the mixture is too weak, it can cause rusting and other undesirable effects. So instead of buying a huge amount of concentrate and mixing your own, see if you can buy a gallon or two of premixed fluid from a local machine shop, enough for the lathe's built-in coolant system if it has one. Be sure your supplier uses a *refractometer* to check concentration—this needs to be spot-on.

3-12 SPRAY BOTTLE INSTEAD OF A COOLANT PUMP

If you don't have a coolant system, use the same fluid in a plastic spray bottle—literally, just like the garden variety. Most shops seem happy to sell premix in small quantities, so don't overbuy. If you keep the premix out of the light in the coolant tank, or in a carefully cleaned 1-gallon bottle (e.g., distilled water), it will be good for a year or more. As it deteriorates beyond the point of usability, you will notice discoloration and oil-water separation, possibly even a bad odor. Evaporate unusable fluid in an open tank; then dispose of the residue as you would any other waste oil (maybe take it to a local auto repair shop).

3-13 LUBING THREAD CUTTING TAPS

A word of caution here: For thread cutting with taps, water-based cutting fluids don't work well at model shop speeds (even though they are used routinely for thread cutting on CNC machines). The better solution is usually one of the "little bottle" cutting oils mentioned earlier. These can be applied drop by drop to the precise location and are thick enough to stay on the spot.

3-14 DEBURRING EDGES OF THE WORKPIECE

No machined part ever leaves the mill in a finished state. There will be sharp edges (burrs), 12 of them on a shape with 6 milled sides, more if there are slots, holes, and other features. Every machinist has a favorite way of dealing with this problem. Some use a deburring tool like that shown in Figure 3-13, with a freely rotating crank-shaped cutter. This can be very effective on long edges, especially sheet material, and is also useful

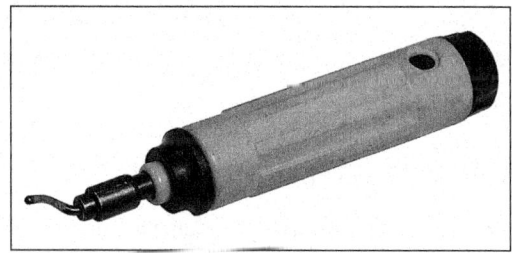

FIGURE 3-13 Deburring tool.

for deburring holes. It takes practice to get results with this tool, which is why many machinists prefer to use a file.

If you have the right kind of file (see Section 3-16, later in this chapter), most milled edges can be smoothed nicely in just two or three stroking passes.

For milled plastic parts there is no need for anything fancier than a single-edge razor blade or a utility knife—a couple of scraping passes should do the job.

Finally, for even faster results on metal, many commercial shops use a belt sander with a very fine grit (or keep a well-worn belt on hand). There are 1" x 30" belt sanders for less than $100 (but be sure you can find 150- or 200-grit belts of the right size).

3-15 DEBURRING DRILLED HOLES

The other finishing task needed on most projects is cleaning up drilled holes—especially threaded holes. Typically, you cannot use a file, because that's fatal to the finished surface. The easy answer, one that works every time, is a pin vise with a countersink bit

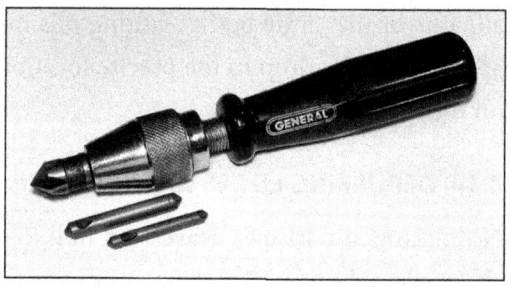

FIGURE 3-14 Pin vise with various countersinks.

(Figure 3-14). I have two pin vises, both 1/4" capacity, with different-size countersinks. (The pin vises are also used occasionally for hole clean-out using a drill or reamer.)

3-16 WHAT TYPE OF FILE?

The short answer is not what you can get from the local hardware store, unless the store happens to stock Swiss pattern files. Standard off-the-shelf files are often good for general shop use, but the mill bastard cut is too coarse for precision work. You might get by with an American second cut or (better) smooth file, but those are not always available. For machine work, I use Grobet files, all #2 cut. (Swiss files are classified by cut number, from 00 to 6, coarsest to finest.) My favorite is the so-called equaling file, 6" long with parallel edges, one edge of which is "safe," no teeth.

3-17 ROUGH-CUTTING METAL WITH A BANDSAW

Every project calls for some means of rough-cutting metal to a size and shape convenient for the mill. You can do this the old-fashioned way, using a hacksaw, but this takes a lot of effort and time. Pretty soon, you will be looking for a power saw. Most small-shop machinists are happy with the 4" x 6" metal-cutting bandsaw sold by importers such as Grizzly and Harbor Freight (Figure 3-15).

FIGURE 3-15 A 4" x 6" bandsaw.

This is a design that has been around for years. It cuts plastics and most metals, including unhardened steel, in sizes up to about 4-1/2" round, and rectangular sections up to 4" x 6".

The blade runs in an oval-shaped frame that swings down onto the workpiece by gravity, counterbalanced by an adjustable spring. The cutting rate is determined by the combination of blade speed (choice of three) and downfeed pressure. Start with low blade pressure (high force from the spring); then work up to the most efficient setting for the material in process.

The 4" x 6" bandsaw usually ships with a 14 teeth-per-inch blade, satisfactory for most work on unhardened metals. Replacement blades are available from McMaster Carr and other suppliers.

Blade tension is important, but you get no help from the instructions. On larger commercial saws, the tension can be metered precisely with an external gauge, but the 4" x 6" saw is not in that league. The best you can do is experiment, putting up with the fact that the blade frequently falls off its wheels—blade tension too low, the blade falls off; too high, the blade falls off. This is something you get used to, and quite soon you have the reinstallation process down to a fine art. (If you look at the geometry, you won't be surprised that the blade falls off—at each end of the very short frame, the blade has to go continuously through a directional change of 45° or so.)

I found over many years with a 4" x 6" saw that it worked best with the blade tight enough to produce a somewhat musical note when twanged, not a dull thud. That's about as specific as it gets. Lubricant on the blade helps preserve the blade, and it also increases cutting speed: Coat both sides of the blade, while it's running, with a waxy lubricant stick. Alternatively, especially for steel, brush on your usual cutting/tapping fluid every 30 seconds or so. The downside, of course, is that lubricants may cause the blade to fall off more readily.

If you have the floor space and budget for it, consider a higher-capacity bandsaw such as shown in Figure 3-16. This saw cuts much faster and more consistently. It comes with a hydraulic feed controller that's a lot easier to adjust than the counterbalance spring on the more basic 4" x 6". Capacity of the model shown is 7" round, rectangular sections up to about 7" x 10".

Sometimes included with saws of this size is a coolant system with a recir-culating pump housed in a 4-gallon tank—essential for high-volume steel cutting but a bit over the top for the small shop; instead, use a spray bottle with Swisslube or other coolant (Section 3-11).

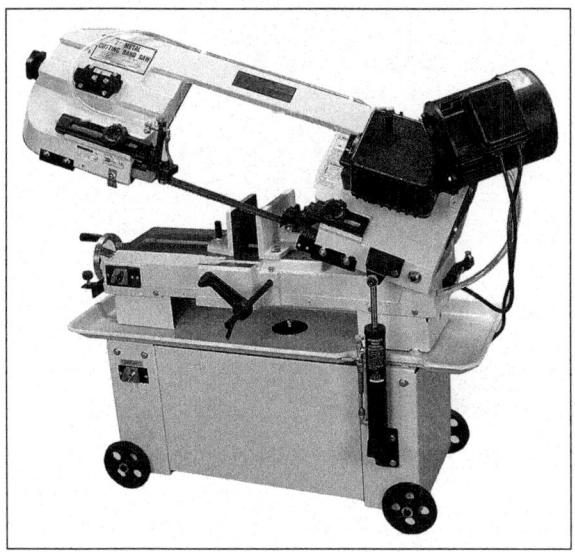

FIGURE 3-16   A 7" x 10" bandsaw.

3-18 BLOCK SQUARING THE EASY WAY

Even with a mill on hand, squaring the surfaces of metal stock is a time-con-suming business. One way to do this is explained in Chapter 6. For every-day work, where precision is less important than appearance and function-ality, a workaround I have used for years is an 8" disc sander made from a salvaged Delta-style belt/disc sander (Figure 3-17). The table it came with was just about usable, but I replaced mine with a beefier steel version that was carefully squared against the disc. I made two fixed-angle slides to go with it, 90° and 45° (for mitering), pinned to 3/4" x 1/4" rectangular bars.

FIGURE 3-17 Modified disc sander. (1) 8" disc, (2) right-angled slide, and (3) mitering slide.

The disc sander is a great time-saver, used to prepare stock on at least 50% of my projects. Highly recommended. It may be that commercially made disc sanders are accurate and rigid enough to do what mine does— or can be shimmed and tweaked to take care of minor discrepancies.

3-19 TIME-SAVING TOOLS

It doesn't take long with any machine tool to wish for a handful of dedicated tools that are always on hand, not put away in the tool chest. Three essential time-savers (Figure 3-18) are:

1. A dedicated socket wrench assembly for the drawbar. The example in Figure 3-18 was made from 1" hex bar, machined square to fit the socket. The tommy bar is 3/8" diameter.
2. A ratcheting combination wrench for vise and other hold-down bolts.

3. A T-handle ball-end hex key, one of three (4 mm, 5 mm, and 6 mm).

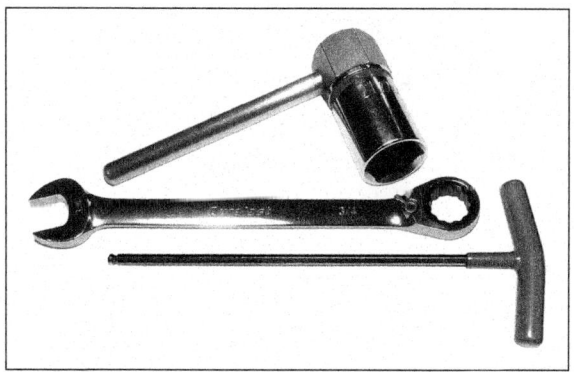

FIGURE 3-18 Time-savers: (1) socket wrench, (2) ratcheting combination wrench, and (3) T-handle ball-end hex key.

3-20 WORKPIECE CLEAN-UP

I'm assuming here that you don't use an air hose to clean off machines and workpieces. Commercial machinists do that in the ordinary course of business, but they are not as troubled as you might be by the effect that can have on the walls of the shop, etc. So we almost always use the less messy shop vac instead. That works for machines and general clean-up, but it usually does nothing for the workpiece.

When cleaning up the workpiece, small-diameter holes (especially blind threaded holes) are often a source of frustration. Try pipe cleaners, of the "scratchy" sort containing brass wire. On steel, use a thin *magnetized* scriber (shown in Figure 3-19) or a dedicated length of thin drill rod (rub it on the surface of a strong magnet).

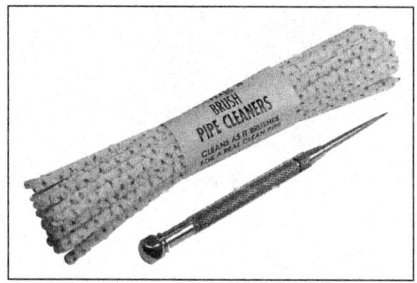

FIGURE 3-19 Two must-have items for clean-up.

3-21 GENERAL CLEAN-UP IN THE SHOP

This is something else you don't want to use an air hose for. The following might strike you as really off-beat, but it can be a minor game changer—no more hauling the shop vac around, tripping over cables and hoses. The setup in my case is nothing more than a hole in the wall at the same height as the inlet port of an ancient shop vac with a standard 2-1/2" hose. The hose, all 21 feet of it, is inside the shop (Figure 3-20).

FIGURE 3-20 Stationary shop vacuum. The power cord runs through the wall to the switched outlet inside the shop.

The vac sits outside, stationary, in a storage area. Power to the vac is from a switched outlet in the shop. When not in use, the hose is simply coiled up on the floor. At one time, it was elegantly coiled on a wall bracket, but the piled-up scheme works better in practice.

This has been a big improvement over previous setups using two or more smaller shop vacs with 1-1/4" hoses. They work passably if cleaned regularly but clog easily on lengthy chips—much more so than the vac now in use. The downside of any machine shop vac is that oil accumulates on every inner surface. Hoses and filters are cleanable using hot water and detergent, but you might want to consider replacing them every year or so instead.

3-22 TWO MORE MUST-HAVES

Finally, two disposable items—flux brushes and paintbrushes (Figure 3-21)—you really can throw away without too much guilt (but you don't have to—instead, you can wash and reuse them for years). I use flux brushes to lubricate machine ways and also to apply cutting oil. Disposable paintbrushes I keep on the bench and on every machine. They are the handiest means I know of for cleaning off drills and cutters before returning them to the drawer—something neglected for years until at last I realized that this really is an overall time-saver.

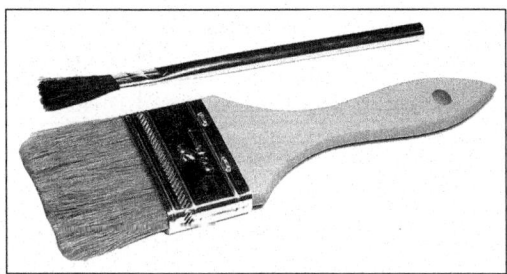

FIGURE 3-21 Multi-use flux brush and paintbrush, both disposable.

HOW HARD IS THAT TOOL?

If you have ever wondered why it is that a file can cut metal off a wood scraper, or even a hardened screw, this may help. These numbers are fairly arbitrary and highly variable, depending on manufacturers' preferences (and quality, which is all over the map—literally).

Typical Rockwell C-scale hardness values

Product	Hardness
Micrograin carbide	75
High-speed steel (HSS)	63–65
Files	65
Case-hardened dowel pins	60
Single-edge razor blades	58
Wood chisels, plane blades	55–66
Hobby knife blades	57–59
Cold chisel	55
Locking pliers	55
Hex wrenches, screwdrivers	50–54
Machinist scales (steel rules)	50
Axes, hatchets	45–55
Socket head cap screws	37–40
Wood scrapers	48–51
Threading tap	60
Hammer face	45–50

3-23 COMMON STEEL ALLOYS

Table 3-1 provides characteristics of a variety of common steel alloys.

Typical chemical composition in % by weight (iron makes up the remaining percentage). *These values are only a basis for comparison.* Expect variability among manufacturers. Trace amounts may be present even when no value (indicated by —) is specified. Phosphorus and sulfur (indicated by *) are regarded as impurities. AISI is the American Iron & Steel Institute.

There are three main categories of stainless steel: austenitic, ferritic, and martensitic. The main difference between *all* stainless steels and simpler alloys, like 1018, is their high chromium content, plus—in some cases—a mix of more exotic elements, such as tantalum and niobium. *Austenitic* steels are the most common. They are nonmagnetic, unless work-hardened. They cannot be hardened by heat treatment. *Ferritic* steels, which are less ductile than austenitic types, also cannot be hardened by heat treatment; notably, ferritic steels are magnetic. *Martensitic* steels, also magnetic, have a higher carbon content and can be hardened and tempered much like non-stainless low-alloy steels.

CAST IRON

Cast iron, sometimes referred to as gray iron, is used for machine tool bases, engine blocks, gears, flywheels, stoves, etc.

Its *carbon* content of 2.5 to 4% by weight is more than 10 times that of common steel alloys. This is the main reason for its very different machining characteristics (powder instead of chips). Other typical constituents of gray iron are *silicon* 1 to 3%, *manganese* 0.15 to 1%, *sulfur* 0.25% max, and *phosphorus* 1% max.

TABLE 3-1   A selection of common steel alloys

STEEL TYPES	AISI ref	Carbon	Manganese	Silicon	Phosphorus	Sulfur	Chromium	Molybdenum	Vanadium	Other
Construction	A36	0.08–0.29%	0.40–1.20%	0.15–0.40%	0.04% max	0.05% max	—	—	—	—
General purpose	1018	0.13–0.20%	0.30–0.90%	0.15–0.30%	0.04% max	0.50% max	—	—	—	—
	1117	0.14–0.20%	1.00–1.30%	—	0.04% max	0.08–0.13%	—	—	—	—
Higher strength	1045	0.43–0.50%	0.60–0.90%	0.15–0.30%	0.04% max	0–0.05%	—	—	—	—
	4140	0.36–0.46%	0.65–1.1%	0.15–0.4%	0.035% max	0.04% max	0.75–1.2%	0.10–0.25%	0.05–0.15%	—
Easy machining	12L14	0.15% max	0.85–1.15%	—	0.04–0.09%	0.26–0.35%	—	—	—	0.15–0.35% lead
	1215	0.09% max	0.75–1.05%	—	0.04–0.09%	0.26–0.35%	—	—	—	—
Tool steels	A2	0.95–1.60%	0–1.00%	0–0.60%	0.030% max	0.030% max	4.75–5.5%	0.70–1.40%	0.15–1.10%	—
	D2	1.40–1.65%	0.60% max	0.30–0.60%	0.030% max	0.030% max	11.00–13.00%	0.5–1.2%	0.5–1.10%	—
	O1	0.85–1.05%	1.00–1.40%	0–0.50%	0.030% max	0.030% max	0.40–0.70%	—	0–0.30%	0.40–0.60% tungsten
	W1	0.95–1.05%	0.10–0.40%	0.10–0.25%	0.025% max	0.025% max	0.15% max	0.10% max	0.10% max	0.15% max. tungsten

STEEL TYPES	AISI ref	Carbon	Manganese	Silicon	Phosphorus	Sulfur	Chromium	Molybdenum	Vanadium	Other
High-speed steels (HSS)	M2	0.78–0.90%	0.15–0.40%	0.20–0.45%	0–0.030%	0–0.030%	3.75–4.5%	4.50–5.50%	1.75–2.20%	5.50–6.75% tungsten
	M42	1.05–1.13%	0.15–0.40%	0.15–0.65%	*—	*—	3.50%–4.25%	9.0%–10.0%	0.95%–1.35%	1.15%–1.85% tungsten 7.75–8.75% cobalt

STAINLESS ALLOYS	AISI ref	Carbon	Manganese	Silicon	Phosphorus	Sulfur	Chromium	Molybdenum	Nickel	Other
Austenitic (typically non-magnetic, unless work-hardened) Not heat-hardenable	303	0.15%	2.0%	1.0%	0.2%	0.015%	17–19%	0.60%	8–10%	—
	304	0.08%	2.0%	1.0%	0.045%	0.030%	18–20%	—	8–10.5%	—
	316	0.08%	2.0%	0.75%	0.045%	0.030%	16–18%	2.0–3.0%	10–14%	0.1% nitrogen
	316L	0.03%	2.0%	0.75%	0.045%	0.030%	16–18%	2.0–3.0%	10–14%	0.1% nitrogen
Martensitic (magnetic, heat-hardenable)	410	0.15%	1.0%	1.0%	0.040%	0.030%	11.5–13.5%	—	—	—
	440C	0.95–1.20%	1.0%	1.0%	0.040%	0.030%	16–18%	0.75%	—	—
	17-4PH	0.07%	1.0%	1.0%	0.040% *	0.030% *	15.0–17.5%	—	3.0–5.0%	3.0–5.0% copper 0.15–0.45% niobium + tantalum

Nominal values only—expect wide variation between suppliers. Trace amounts may be present even when no value (—) is specified. Phosphorus and sulfur (indicated by *) are regarded as impurities. AISI is the American Iron & Steel Institute.

CHAPTER 4

Tramming, Clamping, and Vises

CONTENTS AT A GLANCE

4-1 TILTING THE HEADSTOCK

In routine operations, the user relies on squareness of the spindle relative to both axes of the table. On a bench mill, front-to-back squareness is set at the factory and is not adjustable (by everyday methods), but in the other plane the headstock can usually be set to any angle up to 90° either side of the normal vertical position (Figure 1-15).

Because re-establishing true vertical—tramming—on any mill is a time-consuming process, look first for other ways of handling the project instead of tilting the head.

On a bench mill, the headstock is usually secured by three nuts spaced 120° apart, one underneath and one on either side. A knee mill head-stock has four bolts in a square pattern. The headstock is top-heavy, and it may swing suddenly to either side unless a *helper is on hand* to restrain it. Testing for movability as you go, carefully loosen the nuts by degrees. Be especially careful if the head has not been moved before—the paint seal may let go without warning. (First-time tilting may also call for unusual effort on the wrench.)

Set the headstock to the desired angle by reference to the tilt scale; then retighten the nuts. Bear in mind that the scale is good only to approximately ± 0.25°, so a more accurate means of angle measurement will be needed if the project calls for precise tilting—"tramming."

4-2 TRAMMING

Tramming means accurate alignment—in this case, adjusting the head-stock tilt to bring the spindle to a known angle—usually 90°, "zero tilt"— relative to the table.

As shipped, the mill is set approximately to zero tilt, squared accurately enough for out-of-the-box test drillings, etc. For more demanding project work thereafter, the spindle needs to be set at precisely 90°, in other words *trammed*. "Out of tram" may show up as an offset of a few thousandths between entry and exit of a deep hole, or it may show up as a scalloped effect when surfacing a workpiece (Figure 4-1).

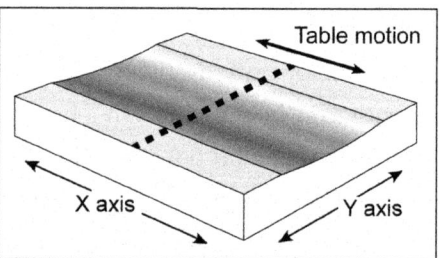

FIGURE 4-1 Head tilt affects surface flatness. This is the effect, much exaggerated, of the head tilted minutely out of square when milling a surface with a large-diameter cutter. The scalloping effect would be at right angles (dashed line) if instead the table is traversed along the Y axis, with the head tilted forward or back (this is not adjustable on bench mills).

Tramming is done by fine-tuning the headstock tilt angle. Tram is typically checked by attaching a dial indicator (see Section 5-10) to some form of "sweepable" holder installed in the spindle, the aim being to adjust tilt for the *same reading* on either side of the X axis.

4-3 HOW TO TRAM THE HEAD

Figure 4-2 shows a typical shop-made holder. It has an offset arbor allowing the choice of two radius arms, a "short" one for tramming the Y axis (front to back of the table) and a "long" one for tramming the X axis. A collet is used to hold the arbor. The dimensions are arbitrary, but the indicator must be firmly attached, and the arm must be rock solid in relation to the indicator spring force (which can be considerable on plunger-type indicators).

A suggested procedure for re-establishing tram includes the following steps:

1. *Disconnect the power.*
2. Set the headstock to the approximate 0° position on the tilt scale; then tighten the nuts enough to avoid unexpected headstock movement.

FIGURE 4-2 Sweeping holder for dial indicator. This example shows a rectangular-section aluminum bar with holes at each end, allowing the choice of two sweep diameters, 6" and 10", measured from spindle centerline to indicator tip. The smaller sweep can be used for front-to-back tramming, also left-to-right tramming as here. The larger the sweep, the more sensitive the tram.

3. Remove the vise, and clean the table surface.

4. Set a 1-2-3 block (or other precision-ground block) on the table under the indicator probe.

5. Switch on the quill DRO, and/or 3-axis DRO if available.

6. Lower the quill using *fine downfeed* to give an indicator reading of about half full-scale. If a knee mill, fully retract the quill and raise the knee. In either case, *avoid locking the quill* (which could fractionally misalign the spindle).

7. Note both the dial indicator and DRO readings; then back off the fine downfeed (or lower the knee) enough to avoid collision when sweeping.

8. If a gear-head or step-pulley mill, select the highest spindle speed (this lessens drag on the spindle, allowing you to hand-sweep the indicator holder more easily from side to side).

9. Reposition the 1-2-3 block to the opposite location on the table.

10. Swing the indicator holder to the new location; then lower the quill, backing off the fine downfeed again (or raise the knee) to give the same dial indicator reading as in step 7.

If the headstock is perfectly trammed—highly unlikely at the first shot—the quill DRO readings (or knee crank dial) should be the same. If not, loosen the nuts just enough to allow the headstock to be "tapped" a fraction of a degree in the direction called for; then retighten the nuts. (The "tap" can be anything from a gentle hand-slap to a rap with a soft-face dead-blow mallet).

Repeat steps 4 through 10 until satisfied with the tram, tightening the nuts as you go. This will likely call for several iterations. There is no "right" tram; the acceptable difference in side-to-side readings depends on the project specs. As a starting point, aim for ± 0.001" on a radius of 5" or 6".

A similar procedure may be used to check tram in the Y axis, front to back. The difference here is that the Y axis tram on *bench mills* is established in manufacture, and it can be adjusted only by shimming (or epoxy-filling) the column-to-base interface. This is a two-person procedure, possibly requiring an engine hoist or some other means of unweighting the column and headstock. (Also keep in mind that the front-to-back tilt can vary as the headstock is raised and lowered.)

On most *knee mills*, front-to-back tilt is easily adjustable in the shop without special methods (some mills have a worm drive to make this easier).

Tramming calls for patience on every mill! Expect to tighten and recheck at least three times (simply tightening the bolts can affect the tram).

4-4 CLAMPING KIT

In most machining operations, the workpiece is firmly clamped to the table, which is moved relative to the cutting tool by the X and Y lead screws. If the workpiece has at least two parallel sides, a vise is the handiest means of clamping, by far—see Section 4-6. Irregular-shaped workpieces take more ingenuity, using various combinations of clamping kit items (Figure 4-3).

FIGURE 4-3 Clamping kit. This is a standard 50-year-old design available from many manufacturers. Quality varies, but they all seem to be functional. Be sure to buy the right size for your table—see the text.

Clamping kits are not one-size-fits-all, so *you need to be careful* when purchasing—for instance, 1/2"-diameter studs are too big for 12-mm T-slots (go with 7/16" or 3/8" instead).

FIGURE 4-4 Using clamps. In this illustration, the step clamps are pivoted on different sizes of step block. The aim here is to place the clamp screws as near to the workpiece as possible to maximize clamping force.

4-5 SQUARING WORKPIECE MATERIAL

Squaring *blocks* and squaring *sheets* of material are very different processes. How to square a block using a vise is described in Chapter 6, Section 6-14. Here we talk about sheets, which are difficult to hold in a vise unless mounted on rigid backing material. Figure 4-5 shows the standard way to deal with sheets of any size within the mill table's dimensions. You will need scrap material at least 1/4" thick under the workpiece to keep milling cutters and drills clear of the table. Safer would be 1/2" or 3/4" ply or (better) MDF, or any other material of uniform thickness.

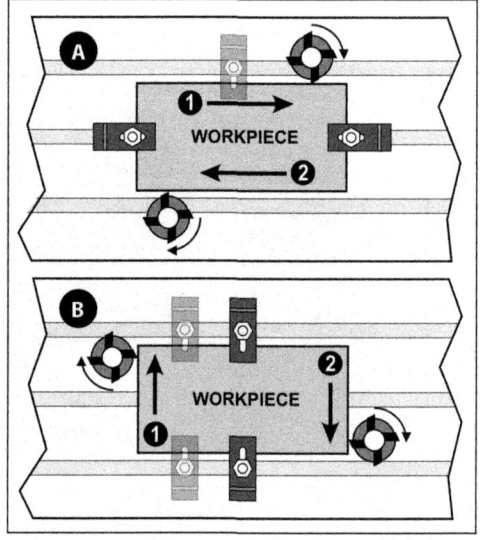

The clamping arrangement in Figure 4-5 is as simple as it gets. Many variations are possible, some calling for additional clamps or clamps in different positions. Cutting the long X axis edge, for instance, would be better controlled with a third clamp positioned as in the ghosted image in diagram A of the figure. With this setup, machine the edge just short of the clamp; then reposition the clamp on the other side of the cutter.

FIGURE 4-5 Machining sheet material. (A) machines the X axis edges, (B) the Y. In this illustration all edges are "conventionally milled," meaning that the cutter at the point of contact is moving in the opposite direction to the workpiece (indicated by arrows 1 and 2). Alternatively, the material could be "climb milled," with the cutter moving in the same direction as the workpiece.

Similarly, the Y axis edges will be held more firmly if the clamps are positioned as ghosted in diagram B.

The key thing here is never to disturb the registration of the workpiece relative to the table; before shifting any clamp, put on another clamp first.

4-6 MILLING VISE

For routine operations the workpiece is usually held in a vise—not just any old machine vise, but one that is made to hold the work precisely parallel with the table in one axis and at right angles to it in the other. The vise is a great time-saver, used so frequently that in many shops it is rarely taken off the table (Figure 4-6).

FIGURE 4-6   Milling vise.

Think of the milling vise as a fundamental part of the milling machine, not just a casual accessory.

Consider stretching the budget to buy the best vise you can. If the price is much lower than say $200 at 2021 prices, it may come up short in accuracy and stability. Vises are usually sized by width of jaw. For small mills, the most suitable size is 4". In my opinion the 5" vise (which is often recommended) is overlarge and can apply an undesirable bending force on the table.

The vise is secured to the table by T-nuts and screws—could be from the clamping kit (shown earlier in Figure 4-3). Some vises come with T-nuts and screws, but there is no guarantee they will fit your mill's T-slots—always a tricky question, because many suppliers are not too sure.

Milling vises also often come with a circular "swivel base" graduated in degrees (Figure 4-7). On the face of it, this looks as though it might be useful, but the question is, for what, exactly? You could imagine drilling a series of holes at a given angular separation (a pitch circle), but that's a job you could do more accurately by calculating a set of X and Y coordinate pairs.

Also, the swivel base is not a substitute for the *rotary table*, which rotates the workpiece in precise angular increments (see Chapter 8). The one type of job I have used a swivel base for is described in Chapter 6.

4-7 MAIN FEATURES OF THE MILLING VISE

FIGURE 4-7 Swivel base.

Precision milling vises are distinguished from garden-variety machine vises by *two main features:* a special pusher device, discussed in this section, and key slots (Section 4-8).

The special pusher device, propelled by the clamp screw, exerts both *forward* and *downward* forces on the movable jaw (Figures 4-8 and 4-9). This was patented by the Kurt Manufacturing company about 50 years ago, and has been incorporated in most—if not all—precision vises since the late 1900s.

This "downward pulling" feature is important because it lessens the tendency of the movable jaw to lift as it tightens on the workpiece.

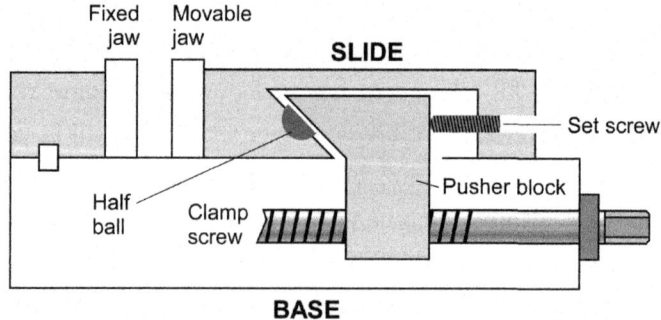

FIGURE 4-8 Milling vise schematic. The inclined surface on the underside of the slide is pocketed for a hemispherical ball, usually hardened. A "pusher block," with a matching inclined surface, propels the slide forward by pressing on the flat surface of the ball. The set screw holds the pusher in loose contact with the ball.

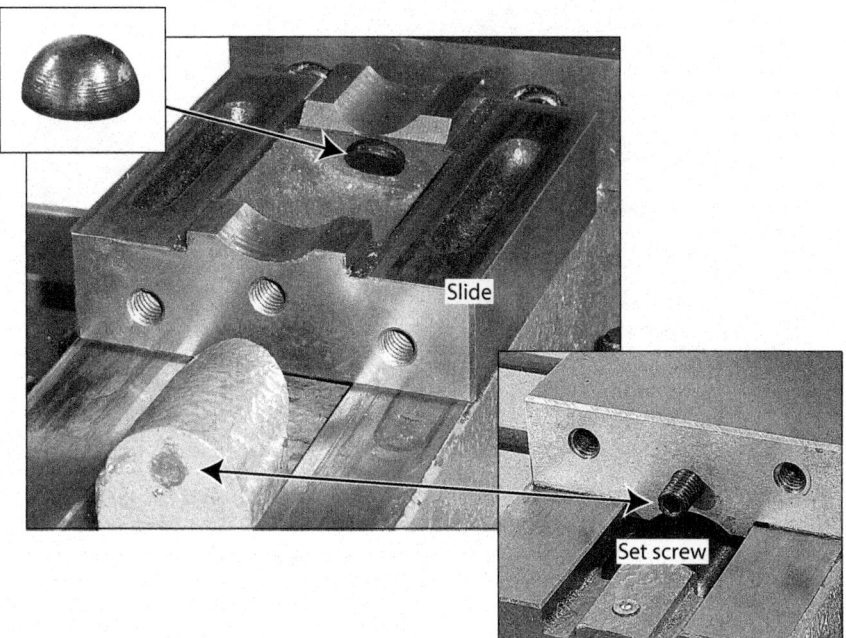

FIGURE 4-9 Vise slide assembly. In the upper view, the slide has been removed and placed upside down on the base. The half ball, shown in the left inset, is held in the slide pocket by grease. The dark impression on the pusher block, indicated by the "double-headed" arrow, is the point of contact of the set screw.

4-8 KEY SLOTS

The second main feature of the precision vise is key slots in the base (Figure 4-10). There are usually two pairs of these, one pair precisely in line with the fixed jaw, the other pair at right angles to it. Key slots can be a great time-saver. With snug-fitting keys they allow the vise to be removed and replaced routinely, accurately enough for general machining *without the need for indicating every time* (see Section 4-11).

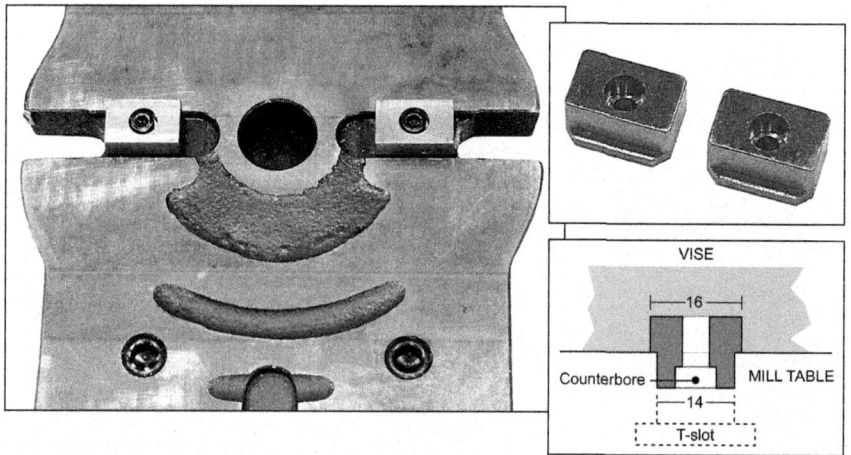

FIGURE 4-10 Key slots with keys. This vise came with 16-mm key slots, the same as the T-slots on larger mills. Most bench mills have 12-mm or 14-mm T-slots, calling for necked-down keys, as shown in the insets. These are not commercially available but can be easily made in the shop.

It is well worth the effort to make vise keys precisely—aim for a snug fit in both vise and table, but not so tight that the vise is unreasonably difficult to lift clear.

4-9 VISE JAWS

The holding capacity of most precision vises can be more than doubled simply by transposing the moving jaw from the back of the slide to the front. For the 4" vise in Figure 4-11, the gripping range increases from 4" to 8-3/4". This works well for substantially flat workpieces that can be situated on the ground-flat upper surface of the slide. Alternatively, the capacity can be increased to 10-3/4" by transposing both jaws—moving the fixed jaw to the very back of the vise.

Transposing the moving jaw takes care of most clamping needs, but once in a while you need vertical gripping higher than the 1-1/2" or so of the standard jaws. "Extended" jaws do not seem to be commercially available for milling vises but are not difficult to make in the shop. You

FIGURE 4-11 Movable jaw repositioned for large work. The movable jaw on most precision vises can be installed on the front surface of the slide, as here, greatly increasing the vise's capacity.

will need 1/2"-thick steel plate, ideally ground for flatness and accuracy (see McMaster Carr, for instance). A 6" x 6" sheet will give you a pair of 4"-wide x 3"-tall jaws (with a useful piece left over). Drill and counterbore (deep, flat-bottomed cavity) for the socket head cap screws, usually M10. Unground steel plate is a much cheaper alternative, but be prepared to do some surface skimming.

4-10 VISE STOP

It doesn't take long with a milling machine to realize that an adjustable stop on the vise would be very helpful when working on a batch of identical parts—just one setup and you're done. *For instance:* Set a stop on piece A, drill holes as needed, replace piece A with piece B, drill holes in the same locations, repeat for piece C, and so on. A versatile stop, shown in Figure 4-12, takes an hour or so to make but can save many hours down the road. The starting point is a 1/2"-diameter rod about 12" long, either cold-rolled steel or (for more predictable diameter) ground drill rod. It is attached to the back of the vise by two M6 x 25 cap screws in the shop-made bushings shown in Figure 4-13 (mine were threaded M10, but that may not apply to all milling vises).

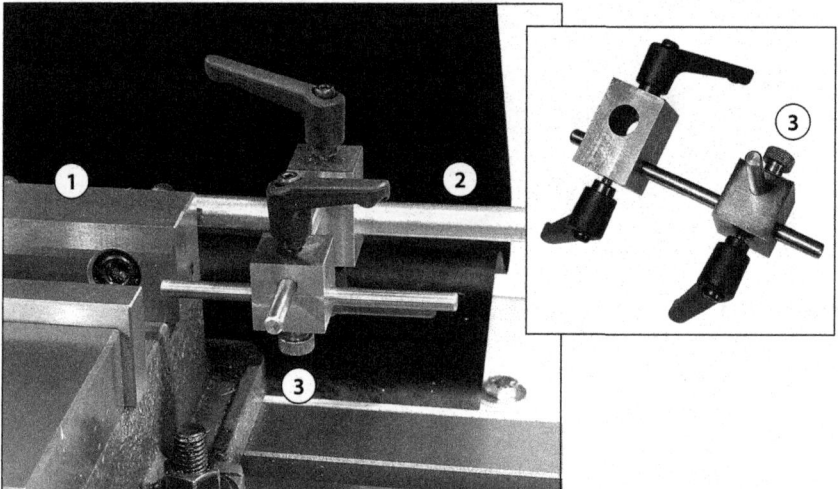

FIGURE 4-12   Stop assembly. Most milling vises have drilled and tapped holes on the back face, either to use for shop-made attachments, as here, or to allow the fixed jaw (1) to be repositioned at the back for extra capacity. This stop assembly slides on a 1/2" bar (2) attached to the two holes on the back face. It uses three M5 ratcheting clamp handles (McMaster Carr 6324K11) for rough setup, plus a brass thumbscrew (3) for final adjustment. Materials: 1"-square steel bar, 1/2" and 1/4"" rod.

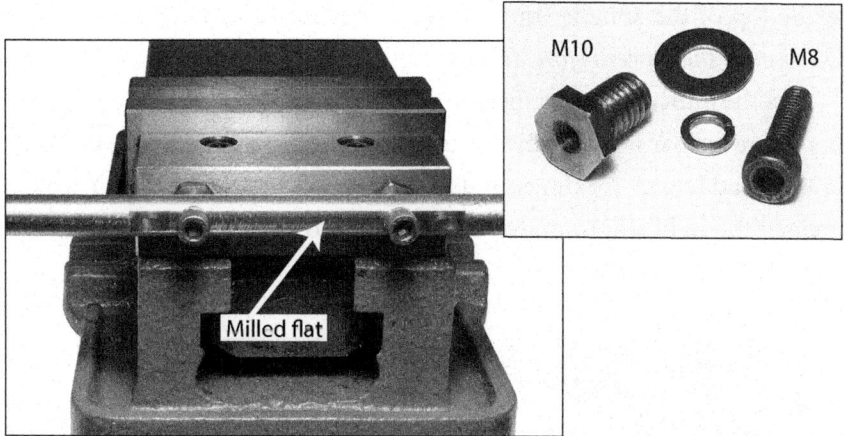

FIGURE 4-13 Vise attachment. The M10-1.5 bushings were turned from hexagonal 12L14 steel, about 3/4" long (the one shown here is 9/16" hex, for the usual reason—found in the scrap pile). *Optional:* Mill the 1/2" bar flat, or counterbore for the socket head cap screws, with lock washers.

4-11 ALIGNING A MILLING VISE

This is a procedure sometimes known as "indicating." In this case, indicating means checking the alignment of the fixed (back) vise jaw relative to the axis of table motion. This is an *essential part* of everyday machining operations. The following assumes you have *not* "keyed" the vise, as described earlier.

Install the T-bolts and align the vise by eye. With one of the clamp nuts lightly snugged, tighten the other one just short of fully tight (but tight enough so the vise won't budge without a definite tap from a dead-blow mallet).

A typical setup for indicating is shown in Figure 4-14. You need to *make sure that the spindle does not rotate* throughout the procedure. If yours is a gear-head machine with no spindle lock, set the gears for the lowest spindle speed. If the mill is belt-driven, it may be necessary to lock the spindle in some other way, such as with a wrench on the drawbar nut at the top of the spindle (or use some other form of shop-made brake). *Disconnect the power before doing this!*

Set the indicator tip against the upper edge of a precision reference bar. If a reference bar is not available, use the front face of the fixed jaw of the vise instead (check for dings, and hone if necessary). Adjust the Y axis to preload the indicator to mid-range at the tightly clamped side of the vise; then lock the Y axis.

FIGURE 4-14 Indicating the vise. The tip of a standard dial indicator, pointed to by the arrow, rides along the face of a precision-ground reference bar. Use the vise jaw itself if a quality reference bar is unavailable.

Note the indicator reading; then watch the indicator as you traverse the table slowly toward the loosely clamped side. Ideally, there should be no discrepancy between the indicator readings at the two ends—unlikely at the first attempt. Return the table to the starting point; then repeat the process, using a soft-face dead-blow mallet to tap the vise from side to side as you go. Repeat the process as often as necessary for the desired accuracy, progressively tightening the "looser" nut. Now fully tighten both nuts, and recheck again (tightening a nut can itself introduce significant error).

An *established routine* like this—tight end to looser end—can save a lot of time. There is no "right" setup for a vise, but as a starting point, aim for an indicator difference of no more than ± 0.001" over the width of the jaw.

There is no guarantee that a vise with keys, described Section 4-8, will indicate satisfactorily. "Perfect indication" is possible only if the slots in the base of the vise are truly parallel to the fixed jaw. Minor misalignment can often be corrected with a metal shim behind the fixed jaw, as shown in Figure 4-15.

FIGURE 4-15 Shimming the vise. Do this if you have installed keys and then found that the vise does not indicate precisely on the back jaw or reference bar. This could be because of manufacturing misalignment of the back jaw relative to the key slots underside.

4-12 WHAT YOU SHOULD EXPECT YOUR VISE TO DO

Figure 4-16 shows why the precision vise is fundamental to efficient milling. A good vise, *one that has been indicated properly*, will do everything shown in the diagrams. You will come to rely on this vise 100% at one point or another in practically every project. A less precise vise will cause endless headaches—random errors of a few mils here and there, unless you fine-tune for good positioning every time you clamp and reclamp the workpiece.

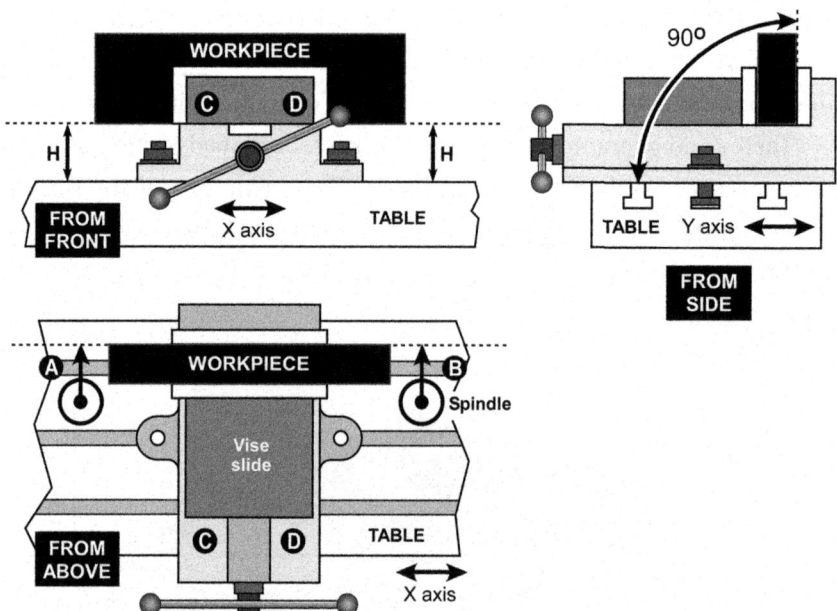

FIGURE 4-16 A quality vise will square the workpiece in all axes. Expect the following: *From the front:* The slide ways C and D should have been ground flat, exactly parallel (H = H) to the underside of the base. *From the side:* The vertical face of the back jaw should be precisely at right angles to the underside of the base. *From above:* The back jaw should be precisely parallel to the table's X axis. This is sometimes established by keys linking the vise to the table's T-slots—see Section 4-8. (If keys are not installed, the vise should be indicated on the back jaw before use, points A and B, and then tightened down for perfect alignment—see Section 4-11.)

4-13 PARALLELS

If the workpiece is at least 1/8" or so taller than the vise-jaw depth, it can be "planted" deep in the vise, with its bottom edge straddling, and resting on, the ground slide ways (C and D in the front view, Figure 4-16). More typically, if the workpiece isn't tall enough to clear the jaws—or if it drops into the gap between the slide ways—it needs to be propped up by "parallels." These are pairs of accurately ground 6"-long x 1/8"-thick steel plates, typically ranging in width from 1/2" through 1-5/8" in 1/8" increments (Figure 4-17).

FIGURE 4-17 Standard parallels.

FIGURE 4-18 Ultra-thin parallels. You can see in this photograph that the widest four pairs, still in wrapping paper after many years, are unusable with my 4" vise. Keeping these thin parallels tight against the vise jaws is always a concern—see Figures 4-19 and 4-20.

Sooner or later, you'll find that the need for a greater working gap suggests a need for "ultra-thin" parallels (Figure 4-18). These usually come in sets of 20 pairs, with widths from 1/2" to 1-11/16" in 1/16" increments. They are ground from spring-steel stock approximately 1/32" thick. The downside of thin parallels, because they are often cut from coiled material, is that they usually don't lie flat. This means you need to take extra care to ensure they are snug to the vise jaws where it matters, well clear of the cutting tool.

One thing to bear in mind about parallels is the amount of space they consume between fixed and moving vise jaws. For instance, a 1/2"-wide workpiece, resting on two 1/8"-thick parallels, has a working gap of only 1/4". This means that the largest through hole you can drill on the centerline is about 7/32"—and that's true only if the parallels are snug to the vise jaws.

Speaking of which, snug contact is always important, and a cause of frustration unless you have a way to hold the parallels apart. Anything

FIGURE 4-19 Parallels separated by springs.

FIGURE 4-20 Parallels separated by packing foam.

springy will do, including assorted springs (Figure 4-19). Another option is a scrap of foam packing material (Figure 4-20) or even a small sponge.

Once in a while, you may have a need for parallels that can be adjusted to specific widths. Figure 4-21 is one of a set of six Starrett parallels rang-

ing from 3/8" to 1/2", 1-3/4" to 2-1/4", length from about 2" to 5". The two parts of each parallel are mated by dovetail ways and are locked with screws. They are thick enough at 1/4" to serve sometimes as a single perch for

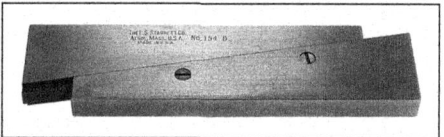

FIGURE 4-21 Adjustable parallel, 15/16" to 1-5/16".

the workpiece, but if that doesn't work, it is often possible to set one parallel and its neighbor in the set to exactly the same width.

4-14 BE KIND TO YOUR MILLING VISE

Milling vises are rugged but not indestructible. One issue with *all* of them is side-to-side looseness of the moving jaw, from 2, 3, and 10 mils and up. This is not a problem, provided that you (1) always regard the mill's fixed jaw as the reference surface, and (2) center the workpiece in the jaws, whenever you can.

But, you may be asking yourself, *What if I need to offset the workpiece to one side, which I often do? What then? The answer:* Do the same as you would with a bench vise—balance it on the other side with a stack of metal packing pieces of the same overall dimension, front to back, as the workpiece.

SHIM STOCK

This is a good point to talk about shim stock, something every shop needs quite often. The usual choices are plastic (vinyl or polyester of some sort), brass, 1018 mild steel, and stainless steel. Sets of various thicknesses ranging from 0.0005" and up are available from a few online suppliers. It's worth shopping around for small sheets if you can find them, say 4" x 6" or 5" x 5". Expect to pay about $10 for a set of plastic shims, a lot more for other materials. If you can't find sets at reasonable prices, buy a sheet or two of say 0.001", 0.002", and 0.005" and double up as necessary. Keep in mind that cut edges of shimstock, unless carefully smoothed, can be unpredictable in thickness—always worth checking with a micrometer.

CHAPTER 5

Other Essentials and First Steps

CONTENTS AT A GLANCE

5-1 R8 SPINDLE FITTINGS

Most general-purpose vertical mills have *R8 spindles*, which means they accept collets, chucks, and other milling accessories conforming to the R8 spec originated by the Bridgeport company many years ago. R8 collets (Figure 5-1) are hardened precision sleeves, bored for various diameters from 1/16" to 7/8" in 1/32" increments (metric R8 collets are available in 1-mm increments from 2 to 20 mm). Collets are used to hold cutting tools such as end mills and reamers more rigidly than a chuck—more accurately, too, provided the tool shank OD is very close to the nominal size of the collet (the closure range is only a few thousandths). The overall length of the collets is about 4". One end of the collet—the business end—is tapered to nest in the mill spindle's internal taper; the other end is internally threaded (7/16"-20) for a *drawbar* inside the spindle. Gripping action comes from closure of the three slits in the collet nose as it is pulled in by the drawbar.

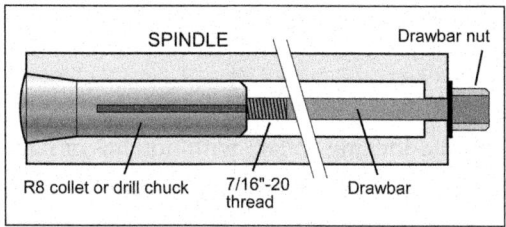

FIGURE 5-1   R8 collets. The keyway, which locates on a pin in the spindle bore, prevents the collet from rotating when being closed by the drawbar.

5-2 INSTALLING A COLLET

First, *cover the table and vise* with ding-resistant scrap material to prevent damage from anything that might fall. Slide the collet up into the mill spindle, feeling as you go for the locating pin in the spindle bore. (*Tip:* To indicate visibly the pin's location, some users *lightly* mark the bottom end of the spindle with a center punch; this can save a lot of frustration, but make sure the dimple can be honed smooth.) Rotate the collet as necessary to engage the pin (or set screw) in the collet keyway; push it in as far as it will go, then screw in the drawbar a few turns to hold the collet in place. Insert the cutting tool into the collet. While holding the spindle lock, if available, fully tighten the drawbar. If there is no spindle lock, and you are working on a gear-head machine, keep the

FIGURE 5-2 Spindle wrench. There are two sizes, 25 mm and 28 mm (1" and 1.1" diameter).

spindle from turning too freely by selecting the lowest speed. Otherwise, hold the spindle at the top using a 6-spline wrench (Figure 5-2). This will probably not be needed on knee mills—they mostly come with a spindle brake, which can serve as a lock when working on the drawbar.

Take care when loosening the drawbar to remove a cutting tool. The tool may drop out unexpectedly, damaging the workpiece, the tool itself, or the table.

Some bench mills have a "spindle-locking collar," with notches or flats, clamped onto the bottom of the spindle. In such cases a dedicated wrench comes with the mill.

5-3 REMOVING A TOOL AND COLLET

Cover the table and vise with scrap material. Unscrew the drawbar a couple of turns; then tap the top of the drawbar with a non-marring hammer such as shown in Figure 5-3. This shop-made example is brass, with a hex-section steel shaft—a useful starter project.

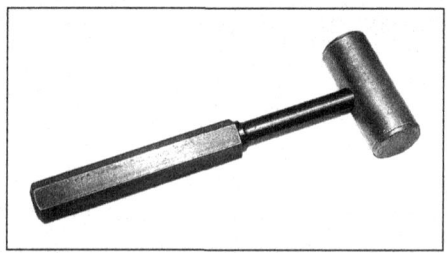

FIGURE 5-3 Use a hammer to free the drawbar.

5-4 DRILL CHUCK

Vertical mills are always part-time drilling machines, so you will need a precision drill chuck. The one-piece style with integral R8 shank (Figure 5-4) is more rugged, and will likely have less runout than the more common variety with a separate arbor. Interestingly, keyless chucks grip drills, etc., almost as reliably as keyed chucks.

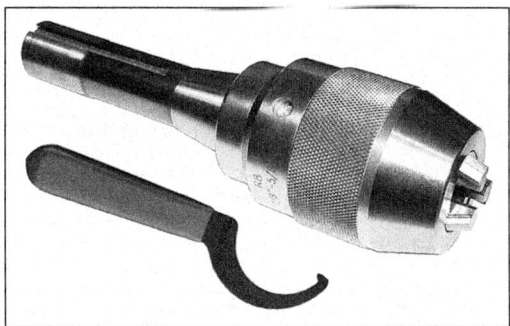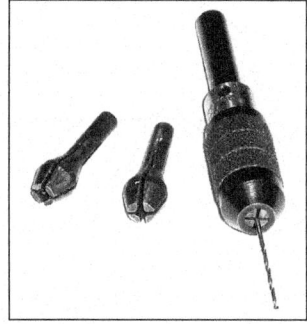

FIGURE 5-4 Typical one-piece precision chuck. Right: Micro drill chuck adapter with three collets for small drills. The drill bit shown here is #60, 0.040" diameter.

Chuck capacity is important, not just at the upper end (usually 1/2" or 5/8"). Many chucks have a lower limit of 1/8", good only down to a #30 drill. A more useful (rare) lower limit is 1/32", good for the entire number drill range. Yet more rare is a genuine zero to 1/2" chuck; these chucks are available, even if you have to go with a two-piece chuck/shank assembly. In this case, the chuck will come with a Jacobs taper cavity, and the separate R8 arbor will have a matching Jacobs male taper. Be sure the tapers are clean, then press the components together. Retract the chuck jaws, then secure the arbor by tapping it gently with a rawhide mallet. A solid assembly is a must—the table could be badly damaged if the chuck parts separate without warning, especially with a heavy cutter installed. To disassemble a two-piece chuck, use chuck removal wedges (available in various sizes from McMaster Carr). You may get by with modified *plastic* construction-type wedges instead; cut U-shaped slots in the ends to insert between chuck and arbor (Figure 5-5).

If your chuck is 1/8" minimum, for small drills you will likely need a micro drill chuck adapter from a watchmaker or model engineer catalog (see Figure 5-4, right). Buy the best you can find—the cheaper ones usually run way out of true.

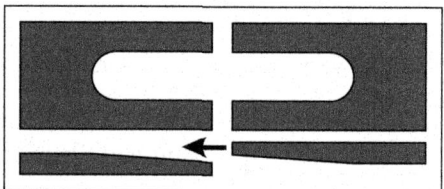

FIGURE 5-5 Makeshift chuck removal wedges.

When installing or removing a chuck, don't let it go unless you're sure the drawbar won't let it drop.

Installing an R8 chuck is exactly the same as installing collets—but there's even more need to *protect the table and vise* from falling objects, *very heavy ones.* With the chuck installed, the mill can be used like a drill press. Simply slide the workpiece around on the table in the usual way. However, that's *not a good idea* unless you have protected the table with a *thick* piece of scrap material, such as MDF, and you have set the quill stop to keep the drill well clear of the table.

5-5 DEDICATED END MILL HOLDER

In many cases, a dedicated end mill holder (Figure 5-6) can be more reliable than a collet, because its set screw pre-

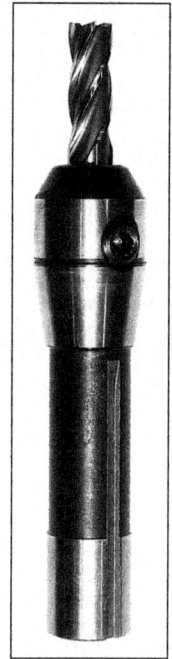

FIGURE 5-6 A 1/2" end mill holder.

vents *axial slippage* of the tool—always a potential problem with a collet, even when fully tightened.

You might see the effect of this when taking a couple of surfacing passes on the workpiece. If the cutter wasn't securely held in the collet, you may be asking yourself the question: *Why is one pass a few mils higher or lower than the other?* The answer is that vibration and cutting forces have shifted the end mill, sometimes to the point where you are cutting a ramp instead of a perfectly true surface. Dovetail cutters, in particular, are very likely to be pulled out of position in this way.

It can save you a lot of this kind of uncertainty if you have end mill holders for the common shank diameters, starting with 3/8", 1/2", and 3/4". One limitation to be aware of: These holders usually work only with *single-ended* cutters, most of which have a crosswise slot for the set screw. To lessen the chances of "downward cutter shift," slightly loosen the screw, wiggle the cutter down as far as it will go, and then retighten.

A POSSIBLE MYSTERY ITEM . . .

Bench mills are often shipped with a shell mill arbor. The example shown here has an R8 shank and a 1"-diameter arbor. It is intended for cutters like the one at the right. It is perfectly functional, but it may not be suitable for use with small machines without the rigidity and/or motor power for cutters 1-1/4" and more in diameter (another reason not to use it is the high cost of shell cutters).

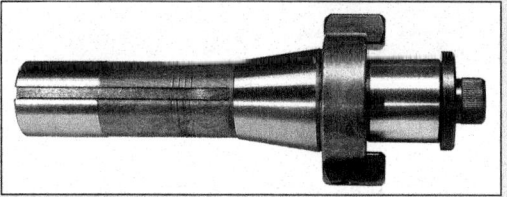

5-6 MILLING CUTTERS

For the small-shop machinist, this is a term that includes three main classes of cutter:

1. Drill bits (Section 5-24)
2. Reamers (Section 5-37)
3. End mills (aka square milling cutters), which are described next

End mills handle about 90% of all work done on a general-purpose milling machine (the other 10% would be drills or special-purpose cutters). This is why it's essential to build, over time, a collection of end mills of various sizes and styles, see Section 5-7.

All end mills have cutting edges on both the underside and the vertical sides; they can therefore machine two surfaces at the same time, as shown in Figure 5-7.

FIGURE 5-7 Shoulder milling: an end mill can cut two surfaces at once.

A special class of end mill, known as "center-cutting," Figure 5-8, has a third function, "plunge-cutting" a cylinder-shaped hole—exactly like a hole drilled with a conventional drill, but with a *flat bottom surface* (hence the term "square" sometimes applied to end mills).

Aside from center-cutting versus non center-cutting, you also need to consider whether the end mill should have two, three, or four flutes (sometimes even six or eight, but only in production operations). For the small shop, by far the most common are 2-flute and 4-flute, either of which are *usable on virtually any workpiece material.* That said, the 2-flute is usually preferred for soft, fast-cutting materials, because its open design allows better chip removal. 4-flute mills, Figure 5-9, are better on hard materials but

FIGURE 5-8 2-flute center-cutting end mill. Most 2-flute end mills are center-cutting, capable of plunge cutting a square-bottomed hole without the need to drill a pilot hole. If you need center cutting, be sure to specify this when purchasing.

have less clearance for chips, and are somewhat better for finishing rather than rough cutting. However, as in most machine shop discussions, there are differences of opinion here, so the choice of end mill style often comes down to personal preferences.

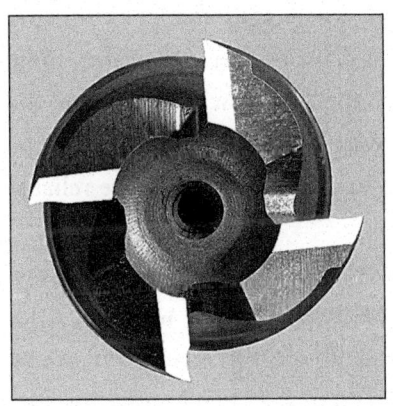

FIGURE 5-9 4-flute end mills. Most 4-flute end mills are *non center-cutting*, left. They cannot be used for plunge cutting without a pilot hole larger than the "dead zone" between the radial cutting edges. Center-cutting 4-flute mills, right, do not have that limitation.

Another factor to be aware of is *helix angle*, the angle between the centerline of the tool and a line tangent to the cutting edge, Figure 5-10. End mills are available with a range of helix angles from less than 30° to 60° or more, but this is important only in production operations where the entire process is fine-tuned for throughput and machining quality. Most small shop machinists don't know—or even have to worry about—what they might find in the end mill drawer, but chances are the mills will mostly have the "standard" helix angle of 30°, which is fine for the vast majority of projects.

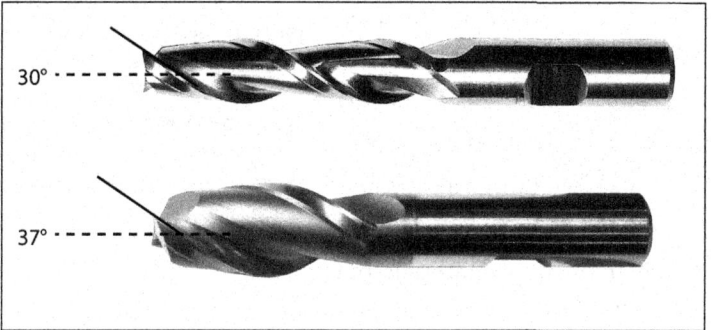

FIGURE 5-10   Helix angle examples.

However, if you are machining a long run of similar components, having converted your mill to CNC, it may be worth the effort of optimizing the helix angle for a specific combination of workpiece material, spindle speed, and cutter feed. The first question then is: How to determine the appropriate helix angle. Answer: Talk directly to the end mill manufacturers, most of whom have good customer support—certainly more reliable in general than blog opinions.

You might even have the most appropriate end mill in stock, but how would you know? (End mills are almost never marked with the helix angle.) There's a simple answer: Wrap a single turn of translucent sticky tape around the tip of the end mill, then use a marker pen to trace the cutting edges onto the tape. Unroll the tape, stick it to a scrap of paper, then measure the helix angle with a protractor.

5-7 BUYING END MILLS

There are literally thousands of end mills and other cutters in the catalogs, in every conceivable variety of sizes, shank diameters, and materials—only a tiny fraction of them are shown in this chapter. If you are really new to all this, just glancing at the handful shown in Figure 5-11 might have you won-

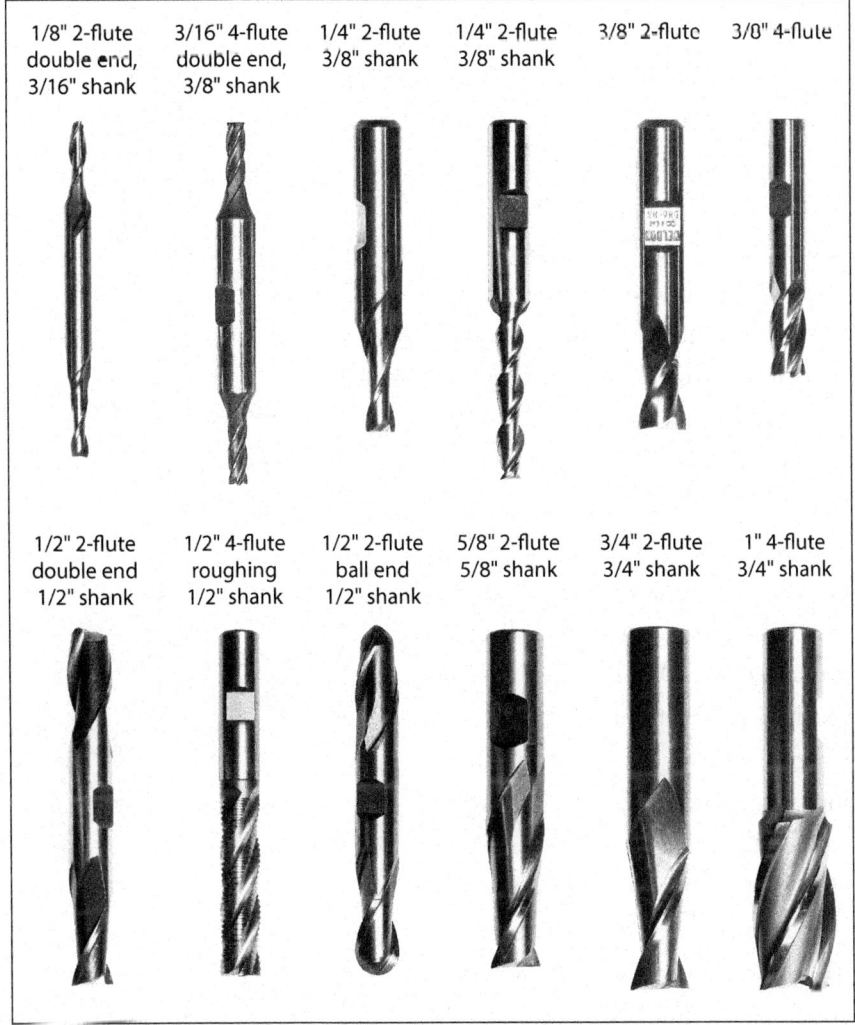

| 1/8" 2-flute double end, 3/16" shank | 3/16" 4-flute double end, 3/8" shank | 1/4" 2-flute 3/8" shank | 1/4" 2-flute 3/8" shank | 3/8" 2-flute | 3/8" 4-flute |

| 1/2" 2-flute double end 1/2" shank | 1/2" 4-flute roughing 1/2" shank | 1/2" 2-flute ball end 1/2" shank | 5/8" 2-flute 5/8" shank | 3/4" 2-flute 3/4" shank | 1" 4-flute 3/4" shank |

FIGURE 5-11 A selection of end mills.

dering where to start. One answer is to buy HSS 2-flutes and 4-flutes in just two or three of the "standard diameters," say 1/4", 3/8", and 1/2", then add to them as and when needed. The 2-flutes, at least, should be center-cutting.

End mills are sometimes offered in single- and double-ended versions. If you are careful always to use the same end every time, double-ends offer twice the useful life. The downside of double-ends, because of their extra length, is that they typically have to be held in collets instead of the more secure R8 end mill holders (Figure 5-6, repeated here in Figure 5-12). End mill *sets* can sometimes be a good buy, but they usually include 1/8" and 3/16" mills, which have a high mortality in first experiments.

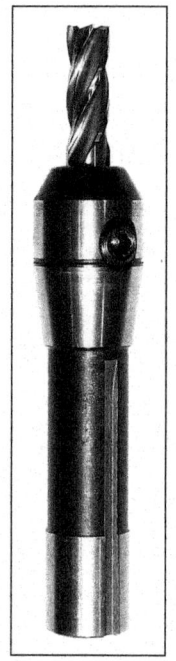

FIGURE 5-12 R8 end mill holder. Use in preference to a collet for most milling with single end cutters.

5-8 OTHER CUTTERS

End mills take care of practically all everyday milling jobs, but there are instances where special cutters are called for. One instance of this is machining a flat surface area that, if larger than, say, 2" x 2" would take a long time and several passes with a 1/2" end mill. This can be done more efficiently with a fly cutter such as that shown in Figure 5-13. One caveat for fly cutting: surface flatness will be an issue if the milling head is not perfectly trammed, see Chapter 4, Figure 4-1.

FIGURE 5-13 Shop-made fly cutter for 3/8"square tools. With the carbide insert tool shown here, the cutter needs to be run in reverse (counter-clockwise looking down on the workpiece).

WHAT MILLING CUTTERS ARE MADE OF

Economy priced end mills today, including imports, are made more or less exclusively of high-speed steel (HSS) alloys, such as M2 and "cobalt steel" M42. No-name generic cutters—unless otherwise specified—are likely to be M2, which is usually fine for non-production operations. The more complex HSS alloys are better able to withstand higher temperatures; but they cost more and are more difficult to grind.

Also available, at considerably higher prices, are solid (tungsten) carbide end mills. These can—should—be run much faster than HSS; also, because they are more rigid and less easily deflected and can deliver accurate dimensioning and better surface finish. The downside of carbide is that it has poor impact resistance. Because of their greater toughness, i.e., resistance to breakage, more exotic powdered metal compounds are today increasingly used instead of tungsten carbide.

Dovetail slots, often used in precision sliding assemblies, typically need a 60° (included angle) cutter such as the one shown in Figure 5-14. Another caveat: dovetail cutters exert a high downward force that tends to pull the cutter out of the collet, changing the depth of cut unexpectedly. Make sure the collet is properly tightened, or (better) use an end mill holder, Figure 5-12.

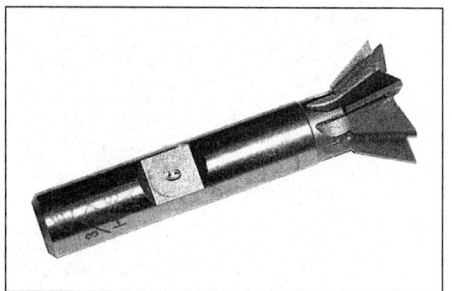

FIGURE 5-14 3/4" x 5/16" dovetail cutter with 3/8" shank.

Next on the list of special cutters is the key seat cutter, Figure 5-15. This is used for Woodruff key slots on circular shafts. Key seat cutters usually have a 1/2" diameter shank with radiused neck. This example cuts a 1/16" slot. For wider key slots, say 1/8" plus, consider using a small end mill instead of a special cutter like this (caveat: to avoid over-sizing the slot, use a 2-flute end mill with high spindle speed and low feed rate).

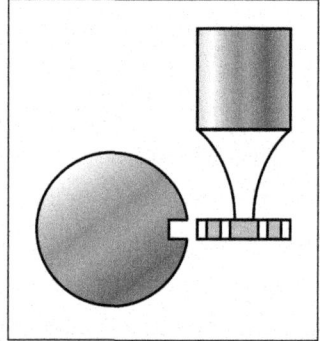

FIGURE 5-15   Cutter for Woodruff keys.

Last of the "other cutters" considered here is the slitting saw, Figures 5-16, 5-17, and 5-18.

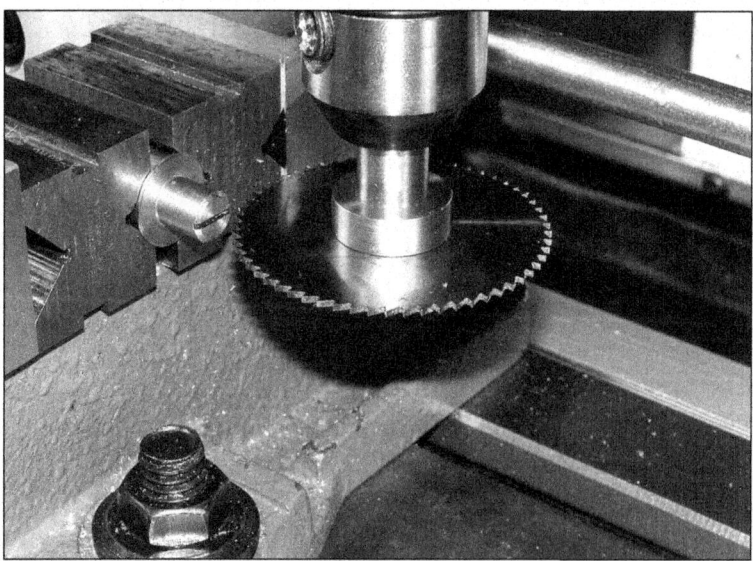

FIGURE 5-16   0.045" thick slitting saw in use.

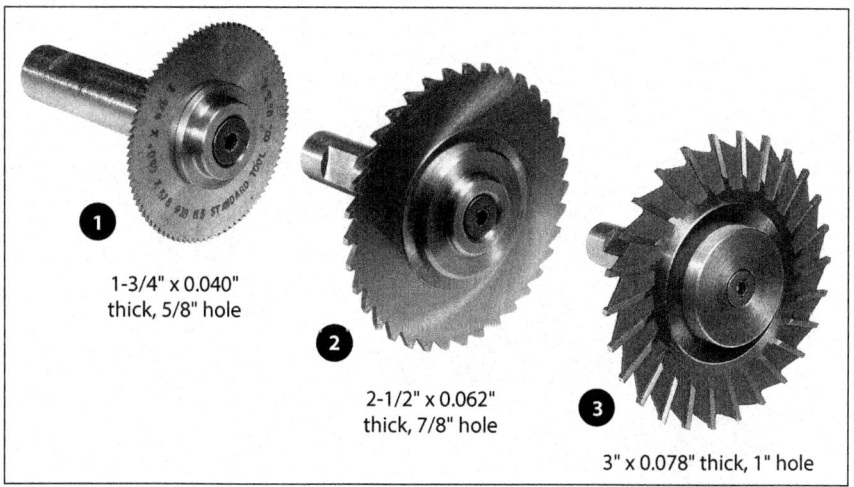

1-3/4" x 0.040" thick, 5/8" hole

2-1/2" x 0.062" thick, 7/8" hole

3" x 0.078" thick, 1" hole

FIGURE 5-17 Slitting saw examples (note the varying diameter clamp plates).

In Figure 5-17, saws (1) and (2) are made from sheet HSS material, usually ground concave; thickness range about 0.01" to 3/16". The more costly style (3) has side cutting edges for better chip removal; thickness range 1/16" to 3/16". One problem with slitting saws is that hole size varies among saws; another issue is cutting depth, which is limited by the mounting hub or clamp plate. There are two ways to deal with problem #1, center hole variability. The easy way is to use an off-the-shelf "universal arbor" with nesting cups of different diameters; these work in a pinch, but they often run so badly out of true that you may opt for a better solution, such as a shop-made custom arbor, which deals with both issues at once, Figure 5-18.

CARE OF MILLING CUTTERS

If you can, reserve your new cutters for brass, aluminum, and plastics. Use them on steel only when they have lost their initial edge. Finally, when they can no longer do a decent job on steel, send them out for regrinding. There are services that will do an excellent job for less than a fourth of the cost of a new cutter, postage included. The only downside is that a reground mill has a smaller-than-nominal diameter—not a real issue in most situations.

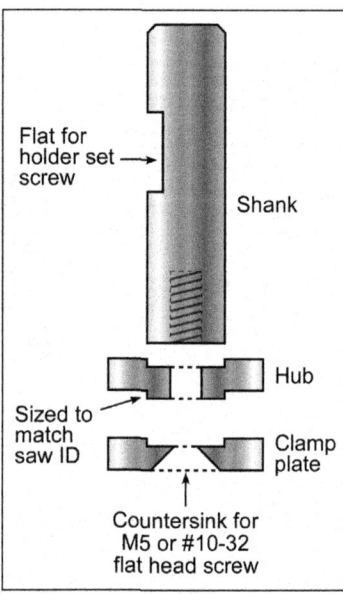

FIGURE 5-18   Shop-made slitting saw arbors. Use 1/2" diameter mild steel bar or (better) drill rod for the shank. To maximize saw-cut depth, make the hub/clamp plate diameters as small as practicable. Use an end mill holder for reliable control of vertical position, Figure 5-12.

5-9 MILLING CHECKLIST

As with practically all machining operations, first experiments with end mills can be disappointing—surface finish bad, dimensions way off, etc. The good news is it gets better. After only a few hours on the milling machine, preventing bad outcomes becomes a matter of instinct—without thinking about it, you automatically attend to an 8-item checklist *every time you change a cutter*:

- Cutter sharpness? (swirl marks on facing cuts, edges look like saw cuts).
- Cutter flexing? (workpiece edge out of square; cut depth and feed rate?).
- Spindle speed optimized?
- Cutting oil needed? (non-ferrous material no; steel, maybe, Chapter 3).

- Milling direction, conventional versus climb? (Chapter 6)
- Table moving smoothly, no rattling? (Gibs properly adjusted?)
- Workpiece "fixturing"? (Vise and/or clamps fully tightened?)
- Cutter firmly held in the spindle? (*All end mills and drills exert a force that can lift the workpiece UP, or the cutter DOWN*)

5-10 DIAL INDICATOR

A dial indicator makes short work of many routine setup operations on the mill, the lathe, and the surface plate. As with most model shop operations, there are always alternative ways to get results, but a dial indicator is definitely on the must-have list. It doesn't have to be a name brand, like the ones shown in Figure 5-19, because in most cases you will be comparing one setting with another, not making precise absolute measurements.

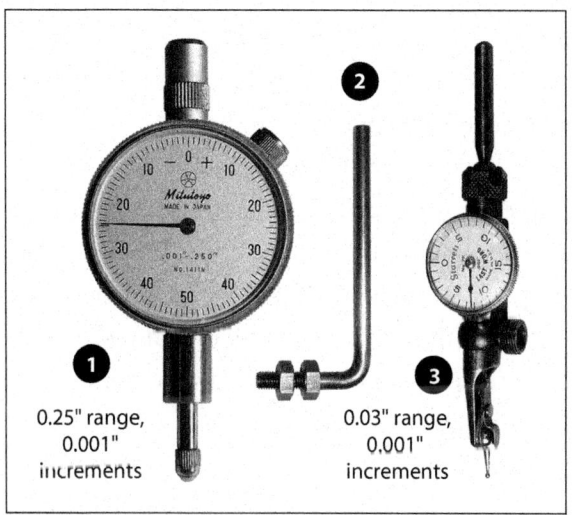

0.25" range, 0.001" increments

0.03" range, 0.001" increments

FIGURE 5-19 Example dial indicators. (1) is the most popular style of dial indicator. Most applications of style (1) call for dedicated holders, such as (2) a shop-made holder for use in a 1/4" collet. Starrett's Last Word indicator (3) comes with a range of attachments for practically all applications, including the 3/16"-round stem shown here. Pricey, but a great time-saver.

5-11 EDGE FINDING: TRADITIONAL METHOD

All precise work on the mill is done with refer-ence to a specific vertical surface on either the machine or the workpiece. There are several types of edge finder—electronic, laser beam, and so on—but the most popular by far are the traditional mechanical finders shown in Figure 5-20. These have been around for 100 years or more, and they are still the edge finders of choice due to their low cost and reliability. The more common version, on the left in the figure, has a 1/2"-diameter shank and a hardened 0.2"-diam-eter tip. The double-ended version, on the right in the figure, is similar, plus it has a cone tip for finding a scribed location on the workpiece (or a center-punched mark or an existing hole).

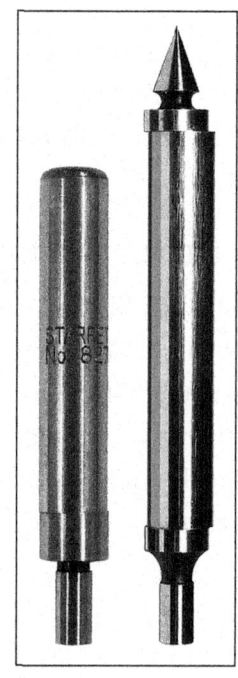

FIGURE 5-20 Edge finders.

To use an edge finder, insert the shank into a chuck or collet; then, with the spindle run-ning at *high speed* (500+ rpm), bring the tip slowly into contact with the vertical edge to be located—suppose, for instance, this is the front-facing surface of the work-piece. Expect to see a noticeable wobble of the small cylinder (if not, push gently on the tip while running). As the Y axis is *slowly* adjusted to bring the workpiece toward the tip, the wobble will diminish to zero, followed by a *sudden snapping action* signifying, precisely, the workpiece location at the moment of full contact.

When this occurs, the spindle centerline on the Y axis is 0.1" from the "found" edge (or if the tip diameter is not 0.2", it will be some other known value). Positioning the centerline exactly on the edge is simply a matter of moving the table handwheel another full turn, in the *same direction it was going* when the indicator "snapped" (this assumes the table lead screw is the usual 10 threads per inch). Even easier (with a DRO) would be zeroing the Y axis, then repositioning by 0.100".

5-12 EDGE FINDING: WIGGLER METHOD

The wiggler is another vintage tool. Like the edge finder in Figure 5-20, it has been around forever and is in regular use by many machinists today. Wiggler kits all seem to be near-clones of the Starrett model in Figure 5-21. The kit contains three probes like the ones shown, plus a crank-shaped mystery item (#5 in the figure). This is not a probe at all: It is an adjustable stem for a dial indicator. It fits the same shank (#1 in the figure) but is strictly a *stationary device*, never powered.

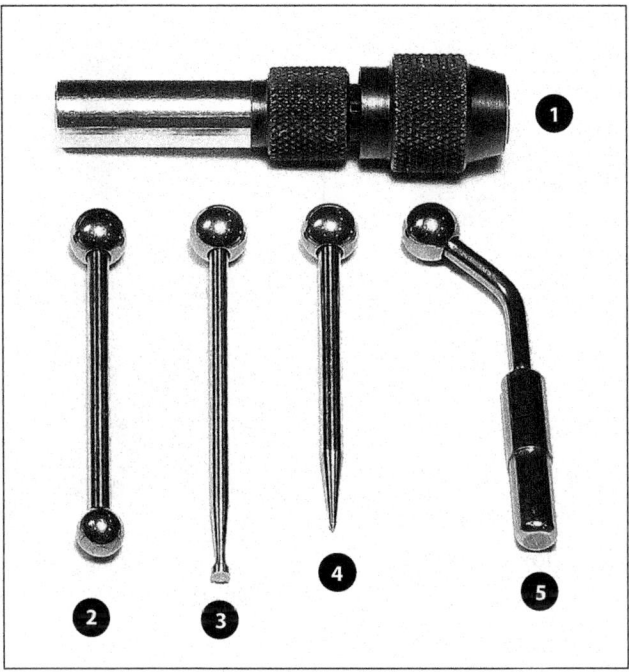

FIGURE 5-21 Wiggler kit. (1) Probe holder with 3/8" shank. A springy collet in the nose allows wiggling motion of the probes (#2, #3, and #4 in the figure), (2) edge-detecting ball probe, diameter 0.250", (3) edge-detecting disc probe, diameter 0.100", (4) point probe, (5) offset stem for dial indicator, stationary use only.

The probe holder is closed by a threaded nose collar, but there should be no need (aside from #5 in Figure 5-21) to tighten it more than *very gently*—the probes should be able to move freely from side to side, with just-detectable resistance, most of which comes from the holder's springy collet. Finding an edge with the ball probe (#2) and disc probe (#3) is described in the following sections.

5-13 EDGE FINDING WITH THE BALL PROBE

As we did in Section 5-11, we'll assume that the vertical edge to be found is a front-facing surface of the workpiece (Figure 5-22). Install the probe in the holder, and then replace the collar. Set the probe *roughly* on the spindle centerline; then run the spindle at high speed (500+ rpm). This will likely cause the probe tip to rotate like a tetherball, a ball on a string. Bring the wiggling under control by gently pressing a wood dowel or pencil against the side of the probe.

Note: Reducing probe runout takes practice: Too much side pressure will cause the wiggling action to flip suddenly from gentle tetherball action to flat-out whirling—no harm done; just raise the quill immediately.

With the probe wiggling just mildly, adjust the Y axis to bring the workpiece slowly toward the probe, watching how the wiggle decays to nothing as the edge is

FIGURE 5-22 Finding an edge with the edge-detecting ball probe (#2 refers to the annotation in Figure 5-21).

approached. At this point, go even more slowly, because the probe will suddenly *flip outward* at the precise moment of full contact. This tells you why the probe ball needs to contact the vertical surface you are trying to "find" as high up as possible, so that when it flips out, it will freely whirl above the workpiece.

You may be wondering, *why do we need the ball to be in the clear when the probe flips out?* Surely, it's not going to damage whatever might be in the way? While it's true that it won't harm a typical workpiece, it *will* wear the probe holder—and the ball inside it—if the motor continues to run while the probe is arrested. Spindles don't stop fast enough, so *raise the quill* quickly instead.

At the moment of "flip-out," the spindle centerline is 0.125" from the found edge (assuming that the ball is 0.250" in diameter). Positioning the centerline exactly on the edge is now a matter of moving the table hand-wheel another full turn, plus 25 divisions, in the *same direction it was going* when the probe flipped. (Or with a DRO installed, you would zero the Y axis and then reposition it by 0.125".)

5-14 LOCATING A SMALL HOLE WITH THE BALL PROBE

To find the exact center of an existing small hole, up to 0.25" diameter, raise the quill so that the ball is just clear of the workpiece (Figure 5-23).

Run the motor, and then apply pressure to the shaft to achieve zero runout. Stop the motor, and lower the quill while adjusting the X and Y axes to center the ball over the hole as accurately as possible. Lower the quill fully to seat the ball in the hole; then raise it clear again. Run the motor. If there is detectable wiggle, the spindle centerline is not exactly over the hole center. If so, stop the motor, readjust the X and Y axes, and then try reseating the ball.

FIGURE 5-23 Finding a hole center (#2 refers to the annotation in Figure 5-21).

5-15 EDGE FINDING WITH THE DISC PROBE

This refers to item #3 in Figure 5-21. In principle, this is exactly the same process as used for the ball probe (Section 5-13), the only difference being that the disc is tiny by comparison, only 0.1" in diameter and 0.05" thick. It can therefore find very small external features and can also be used *inside narrow keyways*.

5-16 FINDING A POINT ON THE WORKPIECE WITH THE POINTER PROBE

Run the motor; then gently apply side pressure on the pointer to achieve zero runout, Figure 5-24. Stop the motor, and lower the quill while adjusting the X and Y axes to position the probe precisely over the desired location (Figure 5-12). A magnifying glass can be a great help for the final adjustment.

FIGURE 5-24 Finding a scribed mark with the pointer probe (#4 refers to the annotation in Figure 5-21).

5-17 HOLDING A DIAL INDICATOR

This has nothing to do with wiggling; it's simply a handy way to hold a dial indicator in a frequently used position. For instance, in Figure 5-25 a Starrett

FIGURE 5-25 Dial indicator setup on wiggler stem (#5 refers to the annotation in Figure 5-21).

Last Word indicator is conveniently mounted for indicating the front and rear surfaces of a workpiece held in the vise (or for indicating the vise itself). For this application, fully tighten the collar, and *don't run the motor.* Disconnect the power to be sure.

5-18 USING AN EDGE FINDER TO CENTER THE SPINDLE ON ANY CIRCULAR FEATURE

If you have a DRO, a conventional edge finder, Figure 5-8, can find the center of a larger hole or a circular rod or disc. Both of these procedures are briefly described in Chapter 7, Section 7-11.

Here is a summary of the basic idea, using as an example a large-diameter rod held vertically in Vee blocks (Figure 5-26). First, note that *the workpiece must be truly circular for a reliable result.*

FIGURE 5-26   Finding the center of a circular rod.

The X axis and Y axis centers are determined separately, as follows:

X Axis

Position the edge finder at the right side of the rod. It doesn't have to be exactly on the centerline of the rod, but the nearer the better. Run the spindle at about 500 rpm. Using the X handwheel only, bring the edge finder into contact with the rod. When the finder "snaps-out," zero the X axis on the DRO.

> **This is important:** Make a note of the DRO's Y *axis* reading!

1. Move the table forward to clear the edge finder. Move the table right, well clear of the finder; then move the table back to the *exact* Y position as previously noted.
2. Move the table left to bring the edge finder again into contact with the rod.
3. When the finder snaps-out, press the X select key on the DRO, followed by the 1/2 key. The X axis will now read zero when the spindle is on the left-to-right centerline of the workpiece.

Y Axis

With the X axis *reading zero*, position the edge finder at the back of the rod. Run the spindle again; then bring the edge finder into contact with the rod, using the Y handwheel.

1. When the finder snaps-out, zero the Y axis on the DRO.
2. Repeat with the edge finder at the front of the rod, X axis *again reading zero*.

3. When the finder snaps-out, press the Y select key, followed by the 1/2 key. The Y axis will now read zero when the spindle is on the front-to-back centerline of the workpiece.

5-19 OTHER WAYS TO ALIGN THE SPINDLE WITH A CIRCULAR FEATURE

Because slight variations in sensitivity of the edge finder can lead to minor inaccuracies, machinists sometimes opt for sweeping around the circular feature with a dial indicator. In most cases, the indicator's support rod is held in a collet, and the spindle is *turned by hand* to sweep the entire circular surface (Figures 5-27 and 5-28). Because dial indicators like this have

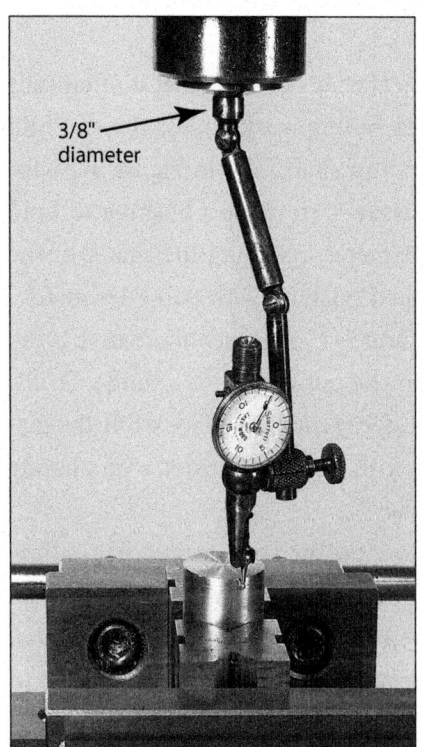

3/8" diameter

FIGURE 5-28 Sweeping the ID of a hole. In this setup the indicator stem fits directly into a 3/16" collet.

FIGURE 5-27 Sweeping the OD of a circular bar. This setup uses a double-jointed rod in a 3/8" collet.

a limited range, it is important to center the spindle approximately before-hand—consider marking the center of the test piece with a fiber tip pen, then use a wiggler with a pointer probe (shown earlier in Figure 5-24).

To make the spindle easier to turn, disengage the drive train (knee mill speed control to neutral), or select the highest spindle speed (bench mill). Start by indicating the surface at the left and right sides, adjusting the X axis for the same indicator reading each side. Set the indicator dial to zero at the midpoint; then rotate the spindle 90° to indicate the front and back. Use a *mirror to view the indicator* when it is facing away. Adjust the Y position for minimum needle movement, which should be close to zero on the dial. Make a few more full-circle sweeps to verify, making minor adjustments as necessary. The spindle is now perfectly in line with the center of the test piece.

A disadvantage of holding the indicator as shown is that it often calls for collet swapping to do the actual machining work, unless the end mill is the same diameter as that of the indicator shaft (3/8" in Figure 5-27, for example). One way to deal with that issue is to make a bushing to hold the indicator shaft, with an outside diameter matching the shank of the end mill. Even better would be a stepped bushing with two ODs matching frequently used end mills, say 1/2" and 3/4". Another alternative is the Indicol, a special frame that clamps on the outside of the spindle. Drills, boring tools, and other cutters can therefore remain in place while the dial indicator is in use. Some Indicols come with a micro-adjuster, a very handy means of fine-tuning the indicator probe.

After using a mirror a few times to view the back-facing dial indica-tor, most machinists will be thinking there has to be a better way. There is, but with the usual downside of expense. The answer is the *coaxial dial indicator*, shown in Figure 5-29, which varies in price from about $100 to several hundred dollars. These indicators are all similar in principle, and all seem to perform reliably. Most of them come with three straight probes of various lengths for sweeping IDs from about 1/8" to 4" and three curved probes for ODs from zero to 4". There is no need for a mirror, because the dial faces front and doesn't move. It is held in place by a restraining arm, which butts against a vertical rod fixed (e.g., magnetically) to the mill table.

The 3/8" shank at the top of the unit is held in a collet, usually *running at 500+ rpm*. Using the coax unit is similar in every respect to the manual sweeping process described above, except that the sweeping is motorized, continuous, and much faster. If your work frequently calls for aligning the spindle, the coaxial dial indicator is another game-changing accessory (just like the DRO).

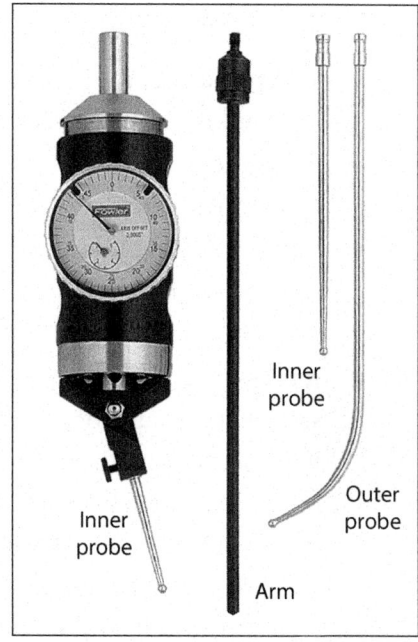

FIGURE 5-29 Coaxial dial indicator. Most versions come with more probes than those shown here.

5-20 DEPTH MEASUREMENT USING A DEPTH MICROMETER

A frequent requirement in mill work is measuring the depth of a hole, slot, or other feature relative to a specific surface. One of the most popular instruments for this is the depth micrometer (Figure 5-30). The measuring range of micrometers like this with a standard-length rod is typically zero to 1". Longer rods are usually available to extend the range in 1" increments. The unit shown measures 0" to 1", 1" to 2", and 2" to 3". Depending on the manufacturer and vintage, the micrometer rod may be either cylindrical or (much to be preferred) flat. A flat rod is less troublesome because it *does not rotate* as the thimble is adjusted. However, to suit the feature being measured, it can be twisted around by hand and thereafter stays put.

Three things to be aware of:

1. Compared with a standard micrometer, graduations on the depth micrometer sleeve and thimble are reversed—which takes getting used to.

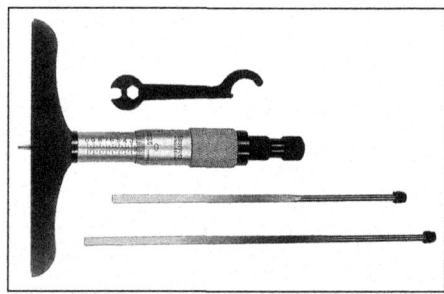

FIGURE 5-30   Depth micrometer, range 0" to 3".

2. It takes practice to apply sufficient pressure for firm contact with the measured feature *without lifting* the base of the micrometer. Use *very gentle* turning pressure on the thimble (the ratchet is usually much too strong).

3. A depth micrometer reads zero when the thimble 0 line coincides with the sleeve baseline. The tip of the 0" to 1" rod should then be exactly level with the underside of the base, both being in firm contact on a flat reference surface. If not, the rod itself has to be adjusted (the sleeve cannot be rotated, as in a regular micrometer). To do this, remove the rod, grip it in a vise (with flat, nonmarring jaws), and then adjust as necessary the locknut at the upper end of the rod—a lengthy, iterative process over two or three cycles of disassembly and reassembly.

To calibrate the longer rods, gauge blocks are needed to hold the micrometer base at a precise height above the reference surface. In Figure 5-31, for instance, the 2" to 3" rod is touching the reference surface, and the micrometer base rests on a 2"-gauge block.

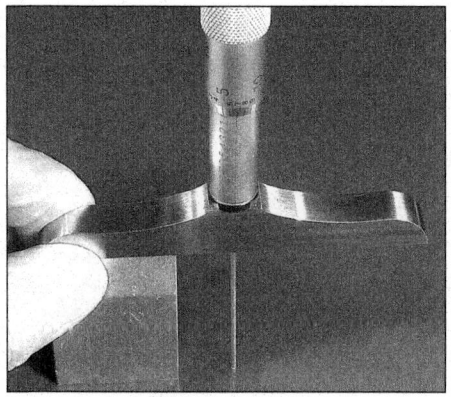

FIGURE 5-31 Calibrate the 2" to 3" range using a 2"-gauge block.

5-21 DEPTH MEASUREMENT USING CALIPERS

The following applies to all slide calipers—vernier, digital, and dial. Many machinists find them easier to use than the depth micrometer, because they are quicker to read, with less potential for error (e.g., base lifting—see Section 5-20, item #2). The dial caliper in Figure 5-32 is my go-to tool for depth measurement, unless greater precision is called for. With the dial caliper, I can usually match depth micrometer readings within ± 0.0015". Another bonus is that it reads depth directly from 0" to 6" with just the one "rod." On the other hand, the width of the blade may be an issue on miniature work—the tip is usually larger than the typical depth micrometer rod. Also bear in mind that the caliper's "base" (the right-hand end of the body) is not wide enough for a reliable stance.

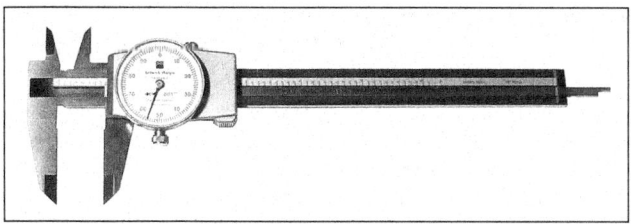

FIGURE 5-32 Dial caliper, range 0" to 6".

5-22 SQUARES

The precision solid squares in Figure 5-33 are very useful in aligning parts at 30°, 45°, and 90° relative to the table and/or vise. These hard-to-find squares are about 0.3" thick, with a long dimension of 3". Standard-pattern 3" and 4" machinists' squares can do part of the job, but you will need some other means of setting the 30° and 45° angles.

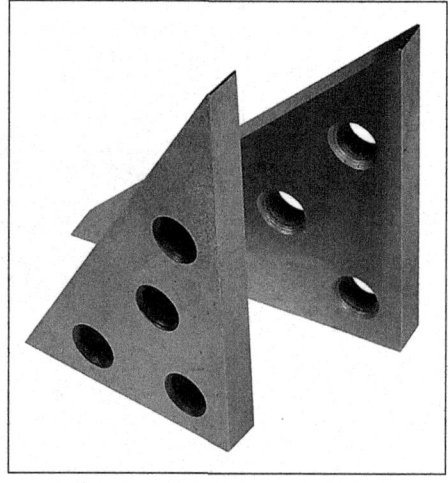

FIGURE 5-33   30-60-90 and 45-45-90 precision squares.

5-23 V-BLOCKS

V-blocks like those shown in Figure 5-34 are used mostly to hold cylindrical workpieces in exact align-ment with the vise. Other sizes are available, with and without clamps, sometimes sold as matched pairs. The clamp hoop in Figure 5-34, right, is within the boundary of the V-block and can be installed and removed with-out touching the vise jaws—a very convenient feature. Smaller V-blocks usually have external hoops, which must sit above the vise jaws; this is not often a problem, provided the block can be raised on parallels.

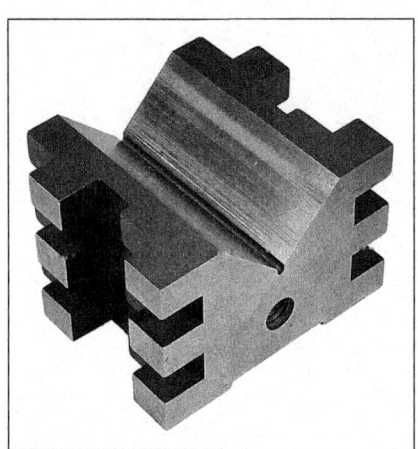

FIGURE 5-34 A 2"-capacity V-block. As shown in the right-hand photo, the clamp screw on this particular model can be installed at 45° (indicated by the arrow) to hold square stock.

5-24 DRILL BITS

There's really no way around it. Every machine shop needs a sizable complement of drill bits, more than you might think. A quick count told me I have about 200 in drill indexes alone, plus dozens of spares and large sizes (Silver & Deming bits). The indexes contain *letter* sizes from A to Z, 0.234" to 0.413"; *fractional* sizes from 1/16" to 1/2" (Figure 5-35); and *wire gauge* (number sizes) from #1 to #60, 0.228" to 0.04" (Figure 5-36).

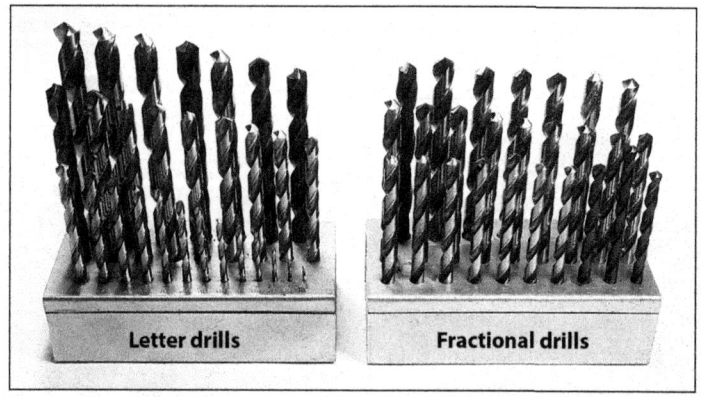

FIGURE 5-35 Letter- and fractional-size drill bits. These are all jobber length.

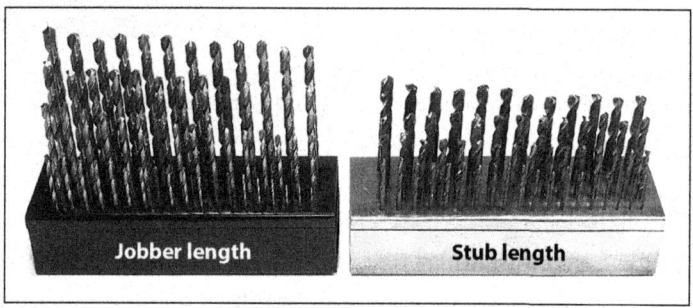

FIGURE 5-36 Number-size drill bits. Stub-length ones are the better choice for most model shop uses.

Drill bits are one of the few items of shop equipment you might want to consider buying in sets. No matter how you buy them, be sure to specify only HSS (high-speed steel). Most of them today are HSS, even the low-price imports, but you need to be sure. (If you don't see "HSS" on the tool shank, it is likely to be ordinary high-carbon tool steel, which doesn't hold an edge reliably, especially if overheated.)

No-name import HSS drills (along with most other HSS cutters) are usually good for model shop use, but they may not hold up in high-volume production.

Sometimes, there are choices of drill point: 135° is better for hard materials (steel, stainless, etc.) than the standard 118°. In practice, there is rarely a need to be fussy about choice of point angle. Both will work fine with practically any material.

Drills are often offered with various coatings for longer life, faster cutting, etc. At the model shop level, it's difficult to see the difference, and it probably doesn't matter enough for a second thought. The drills shown in Figures 5-35 and 5-36 are all either uncoated steel or black oxide coated to provide a degree of self-lubrication (unlike more exotic coatings, black oxide comes with no significant cost increase).

5-25 DRILL LENGTH

The standard-length drill bit stocked by hardware stores is known as "jobber length." (Who knows why?) More useful for the model shop, especially in small diameters, are stub-length drills (aka screw-machine length). These shorter drills flex much less, are less likely to drift off-center and can often deliver good positioning accuracy without the need for spot-drilling with a center drill. My jobber-length drills are used only for deep holes, so they gather dust most of the time.

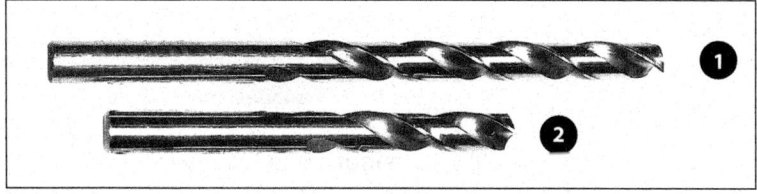

FIGURE 5-37 Jobber- and stub-length bits. These are both 1/4" diameter: (1) Standard (jobber) length 4", (2) screw-machine (stub) length 2-1/2".

5-26 DRILL GEOMETRY

There are two main classes of drills used in the machine shop, differentiated by *point angle*, usually 118° or 135°. General-purpose jobber-length drill bits from the hardware store almost always have a 118° point angle (Figure 5-38).

Both 118° and 135° drills can be used on practically all materials, but the general rule is that the sharper the angle, the better it is for softer materials. Many machinists prefer the "flatter" 135° point angle, which is said to deliver a more truly round hole that's closer to nominal size. That could be so, but in regard to hole diameter, I have never been able to tell the difference between brand-new 118° and 135° bits when drilling mild steel.

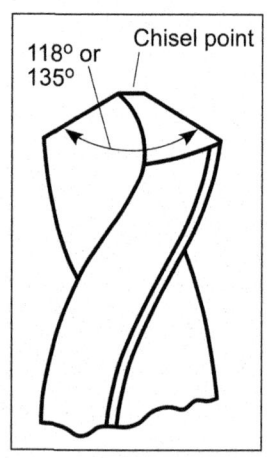

FIGURE 5-38   Drill point angles.

However, if you have 118° and 135° bits of the same size and quality on hand, you will definitely notice a difference between the two when drilling a tough material such as stainless steel. The 135° drill cuts to diameter faster and is less apt to walk off-center.

But no matter what the material or the point angle, it is good practice to "spot-drill" a pilot dimple beforehand. For this, you will need *center drills*, also called "drill-point counter-sinks," with body diameters of 1/8", 3/16", 1/4", and 5/16", trade sizes 0–4 in Table 1. Figure 5-39 shows a fifth trade size with a diameter of 7/16"—but you may not need to go up that high.

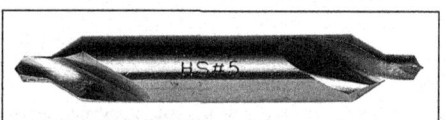

FIGURE 5-39   Number 5 center drill, 7/16" shank, 3/16" tips.

Though these double-ended 60° bits are classified by "trade size," this is not a complete description, because each trade size can be special-ordered in various lengths (the smallest sizes, 0, 00, etc., also come with various tip diameters).

These are the same drills used to prepare material for center-turning on the lathe.

TABLE 5-1 Most common sizes of countersink bits for the small shop

Trade Size	Body Diameter	Drill Tip Diameter
0	1/8"	1/32" (0.031)
1	1/8"	3/64" (0.047)
2	3/16"	5/64" (0.078)
3	1/4"	7/64" (0.109)
4	5/16"	1/8" (0.125)

5-27 SPLIT-POINT DRILLS

This is one more complication that applies mostly to 135° point angle bits. Until recently, 135° bits looked just like the general-purpose 118° variety, but with a less pointy tip. Today, all 135° drills, even the smallest wire gauge size (#60, 0.04"), are available with split points (but not all suppliers stock them—you need to ask). Split-point drills (Figure 5-40) cut with about 50% less thrust than standard drills and also stay on point more reliably. *What's the downside? The answer:* Sharpening.

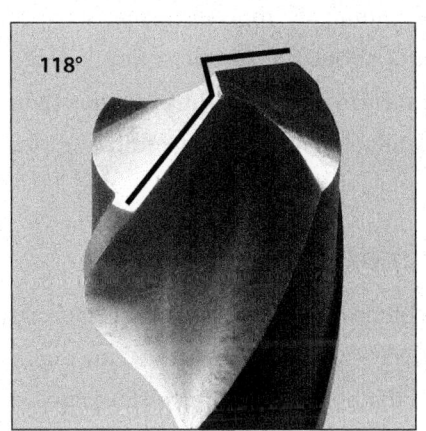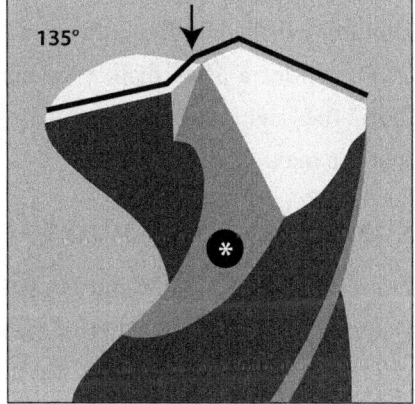

FIGURE 5-40   Drill-point geometry comparison. The 118° drill (*left*) has a standard chisel point. On the 135° drill (*right*), the chisel point has been split by two secondary grinding operations—the surface shown here, indicated by the asterisk, and its counterpart behind the drill. Cutting edges are outlined in black. There are three cutting edges on the 118° drill, four on the 135° drill. The split point, indicated by the arrow, forms a better-defined, smaller center pivot.

Split points are out of range even for the most skillful toolmaker to grind by hand (yes, people still do that sometimes on regular drills). Doing it right takes a dedicated machine with a fine-grit diamond wheel. I don't have such a thing, so I send drills out to a sharpening service. Either that, or I buy new if the cost is less than regrinding.

5-28 DRILLING DEEP HOLES OF ANY SIZE

The deeper the hole, the greater the accumulation of chips in the flutes. This calls for frequent withdrawal of the drill for cleaning and lubing. When drilling deep holes in plastics that tend to squeeze the drill, use progressively smaller drills as you go deeper. When you are to depth with a small drill, finish the hole with the right-size drill.

5-29 DRILLING SMALL HOLES

For less flexing under pressure, use stub-length drills if you have them. If working on steel, back the drill out frequently for *lubing*. Run the spindle faster than you might think reasonable.

Example: According to *Machinery's Handbook*, mild steels such as 1018 should be drilled with a cutting speed of about 100 feet per minute; that's 2000 rpm for a 3/16" drill, 1500 rpm for 1/4". For anything smaller than 3/16", the "right" speed is simply not available on the average mill—so run it at the max.

5-30 DRILLING LARGE HOLES—*CAUTION*

Don't try to drill them in one go. Bad things happen when a lot of pressure is applied to a drill bit. Anything above 1/4" goes better with a full-depth pilot hole, about one-third of the final diameter. Use a *much slower speed* when enlarging a hole.

Example: Drill a 1/4" hole at the regular speed; reduce to *one-half the speed or less* when enlarging to 3/8" or 1/2".

How do you stop large drills from slipping in the chuck? Not easily, but don't fall for the usual remedy of overtightening (and damaging) the chuck.

The only legitimate solution is to work up to the final diameter using progressively larger drills. It takes more time than we like, but it works every time.

Be wary of marketing claims of a mill's drilling capacity—for instance, 1" diameter in steel. This probably means that the motor is powerful enough to make chips with a drill of that size. But that isn't something you need to try (see above). Also note that on small bench mills, you will probably see the headstock wobble when drilling large-diameter holes. This is alarming at first sight, but it usually doesn't cause permanent damage. Tightening the headstock gib strip against the column dovetail may help a little, but don't overdo this. Again, the best fix is to work up to the final diameter in small steps.

5-31 DEBURRING A DRILLED HOLE

Use a standard 82° or 90° countersink bit with 1/4" shank held in an adjustable pin vise. This is actually a keyless chuck mounted on a handle (see Chapter 3, Figure 3-14). For larger holes, I use a 5/8"-diameter countersink bit, which is beefy enough to be handheld without a separate handle.

5-32 OVERSIZED DRILLS

Reduced-shank drills, aka Silver & Deming drills, are available in sizes up to 1-1/2" diameter, with 1/2" shanks (Figure 5-41). Round shanks can slip in the chuck, so choose flatted shanks if you can find them (three flats at 120°). If your Silver & Deming drills have round shanks, consider using a collet instead of a drill chuck.

> **A word of caution:** Large drills like these put a heavy load on the mill and the workpiece. With large drills—regular style or Silver & Deming—the basic rule is to run the spindle at a *very low speed*, at least to start with.

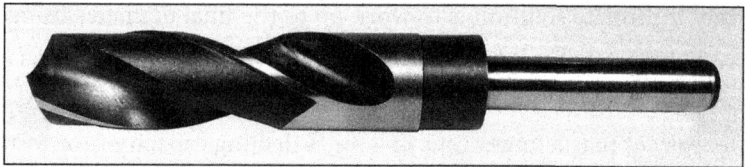

FIGURE 5-41   A 1" Silver & Deming drill with 1/2" shank.

5-33 DRILLING A FLAT-BOTTOMED HOLE

Predrill close to the final size in the ordinary way; then finish with an end mill. It is not necessary to use a center-cutting end mill provided the pre-drilled hole is only a little smaller than the end mill diameter.

5-34 CONTROLLING DRILL DEPTH—BENCH MILL

Most bench mills have a quill DRO, which can be set to zero when the outer cutting edges of the drill are even with the surface of the workpiece. To dry-run the drilling action, raise the quill; then move the workpiece clear. Lower the quill for the desired drill depth on the DRO; then set the quill depth stop—if available. Unfortunately, many of the smaller bench mills, as shipped, do not have a depth stop (but some users add their own shop-made versions). On the other side of the coin, the quill DRO can monitor *drill depth in real time*. Additionally, the small bench mill will likely have a *fine downfeed handwheel*, which can be used instead of the regular "drill press lever." The advantage of the handwheel is much better control of quill depth.

Without a depth stop, there is a degree of uncertainty!

This uncertainty is due to backlash in the drive between the handwheel and quill. With *small drill bits*, this is not likely to be an issue, meaning that the downfeed will cease predictably at the indicated depth. However, this may not be so with larger drills, say 3/8" and up: There is the possibility that the drill will dig into the workpiece, pulling the drill past the desired end point. This is because the pulling force from the drill has overcome the quill return spring, adding say 30 mils to the depth. The workaround for this is to arrive at the final hole size using a progression of drills, increasing the diameter by only a few mils each time. The aim of this is to leave very little metal for the next drill to grab onto.

Most of the larger bench mills have both a depth stop and a quill DRO (Figure 5-42). To set the depth stop, use the fine downfeed handwheel to run the quill down to the desired depth; then clamp the quill. Rotate the depth stop thumbwheel (indicated by the arrow) to lower the stop nut as far as it will go. Release the quill clamp, and disengage the fine handwheel; then, using the drill press lever, lower the quill to the depth stop a few times to verify its setting. Adjust if necessary.

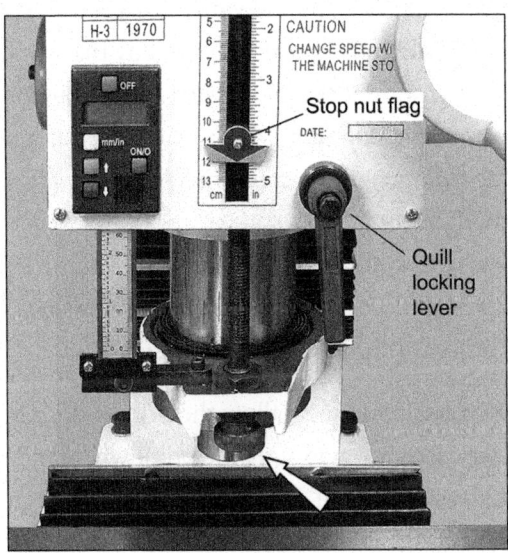

FIGURE 5-42 Depth stop on a large bench mill.

Something else to consider: An end mill has less "grabbing power" than a twist drill and is therefore a good way of drilling to depth if you have a cutter of the desired final diameter.

If your mill has a 3-axis DRO, you may be thinking that an alternative to depth stop issues might be to lock the quill, then lower the headstock (if a bench mill), or raise the knee. You *can* do this, but the downsides are: (1) there is no feel for how the drill is working, and (2) smooth headstock motion on a bench mill can be doubtful. Knee mill users have a better time of it, because the knee can be raised predictably in very small increments (but even so, there's the danger of overshooting by a mil or two). Experiment with scrap material and various drills before working on a project item.

5-35 CONTROLLING DRILL DEPTH—KNEE MILL

Knee mills have no quill DRO, but they do have a better depth stop mechanism (Figure 5-43). Quill position is indicated by the *stop collar* and the 5" scale alongside. Downward travel of the stop collar—and also the quill—is limited by a *micrometer stop nut*, which can be set at any height on the depth screw. One full turn of the micrometer usually gives a vertical shift of 0.05" (= 50 divisions of 0.001").

One way to use this system is to have on hand shop-made spacer blocks or rods of various heights corresponding to the *drill depths* you are likely to need. A typical procedure would be as follows:

1. With the spindle running, allow the drill to cut into the workpiece, stopping at exactly the point where the cavity has expanded exactly to full diameter, as shown in Figure 5-43 (1).
2. Clamp the quill.
3. Place a spacer block of the desired drill depth—say 2"—between the stop collar and the stop nut; then, run the stop

nut up the depth screw to touch the underside of the spacer, as shown in Figure 5-43 (2).

4. Lock the stop nut in place with the clamp nut.
5. Remove the spacer block; then unclamp the quill.
6. Drill the hole to depth.

Suppose that over time you have accumulated a set of spacers for commonly encountered drill depths, but now there's a need for a depth not in the set. The usual workaround for this is to install your most closely matched spacer on the short side; then make up the difference using feeler gauges or other known-thickness packing.

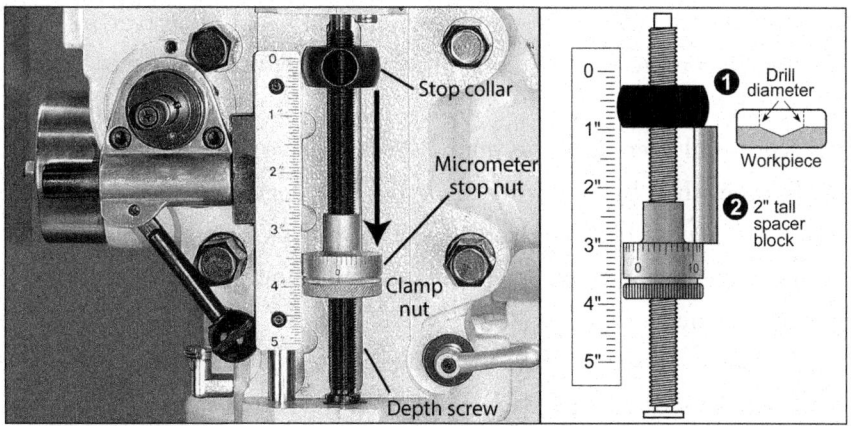

FIGURE 5-43 Depth stop on a knee mill. The diagram, right, shows how to set the stop for a specific depth of travel (2" in this example).

5-36 "PECK DRILLING" TO DEPTH

The following applies to mills of every size and type, when the drill size and/or material are beyond the machine's "comfortable capacity" in terms of rigidity and motor power.

Peck drilling (step drilling) means taking a series of small, controlled-depth nibbles to avoid the problem of overdeep drilling that can suddenly occur when a large drill digs into the workpiece. The chances of this happening are *reduced*, not eliminated, by the depth stop on larger bench mills and all knee mills. Why the word is "reduced" is because the dig-in can be powerful enough to damage both the workpiece and the machine.

How deep to make each step in the series is impossible to generalize, but a good number to start with is 0.01". On a bench mill, zero the quill DRO at the point where the drill has just completed a full-diameter conical cut, as in Figure 5-43 (1). Run the stop nut down as far as it will go (usually against the headstock casing), and then clamp the quill. Raise the stop nut by 0.01" (use a feeler gauge), unclamp the quill, and then drill down to the stop. Repeat as many times as necessary to arrive at the desired depth, as reported by the DRO. This sounds tedious, and so it is; however, it's often possible to confine the "10-mil-a-time" procedure only to the last quarter of an inch or so.

Knee mill users have no quill DRO and have no way to gauge drill depth other than by observing the stop collar against the 5" scale. If the mill has a 3-axis DRO, there are several ingenious ways to achieve better accuracy. For instance, you could lock the quill and then drill to depth by raising the knee.

If, like most machinists, you would not be 100% happy with that method, consider this approach (it assumes you have a spacer equal to the desired drill depth):

1. Fully retract the quill; then drill a full-diameter conical cavity Figure 5-43 (1) by raising the knee.
2. Zero the DRO Z axis; then use the drill press lever to drill *close* to the final depth, estimating from the scale (Z axis still zeroed).
3. Fully retract the quill again; then place a depth spacer between the stop collar and the micrometer stop nut, as in Figure 5-43 (2).
4. Run the stop nut up to the spacer; then remove the spacer.

5. Raise the stop nut another 0.1"—usually two full turns of the micrometer dial—and then drill down to the stop.

6. Back the stop down by a few mils; then drill again.

Repeat the process as necessary to achieve the desired depth, checking repeatedly with the spacer—and progressively reducing the depth increments when nearing the end point.

5-37 REAMERS

For an accurately sized hole, with very little effort, nothing beats a reamer (full name "chucking reamer"), shown in Figure 5-44. These come in two styles: straight flute as shown and, less prevalent, helical (spiral). There is not much to be said about the relative merits of the two styles, but one thing is certain—if the hole to be reamed already has a key slot or cross-hole, spiral flutes are preferable.

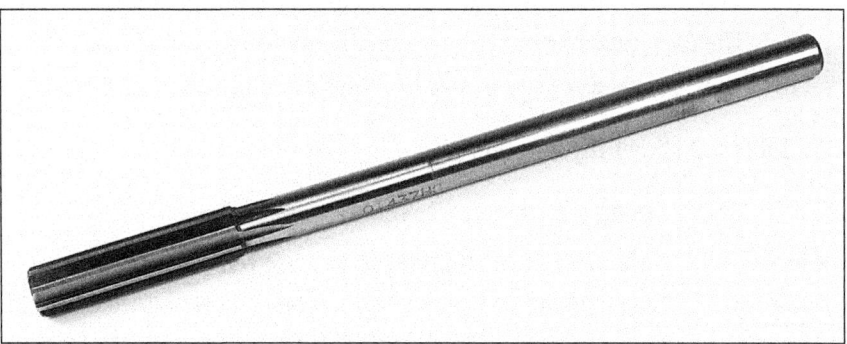

FIGURE 5-44 Chucking reamer.

Reamers are used to enlarge precisely a hole that's a few thousandths undersized—no more than 0.005" on the diameter for holes between 3/16" and 1/2", even less for smaller holes. If you leave more material, the reamer may be pushed out of line, oversizing the bore. The other thing to bear in mind is *spindle speed*, especially when reaming steel—about one-half of the speed you used for drilling—and be sure the reamer is lubed to lengthen tool life.

Golden rules for reaming:

- Run the spindle *slower* than the drilling speed.
- Lube thoroughly.
- Leave only a small amount of material for reaming.
- Don't push a reamer hard against the bottom of a blind hole—it will cut oversize if you do.
- Withdraw the reamer slowly, spindle still turning.

Not knowing how fortunate I was at the time, many years ago I acquired from a retiring toolmaker two sets of straight-flute reamers, 14 sizes each set. The reamers came in index boxes, one marked "Dowel Pin" and the other "Over/Under":

- The *dowel pin* reamers are used to size holes for dowel pins when you need them to be really tight-fitting (so tight *they will actually distort the work* if you don't pay attention).
- The *over/under* reamers have equally obvious applications. They are used mostly to size bearings for standard-size shafts.

Table 5-2 shows my two sets.

TABLE 5-2   Dowel pin and over/under reamer sets

Nominal	1/8 0.125		3/16 0.1875		1/4 0.25		5/16 0.3125	
Dowel pin	0.123	0.1247	0.1855	0.187	0.248	0.2495	0.3105	0.312
Over/under	0.124	0.126	0.1865	0.1885	0.249	0.251	0.3115	0.3135

Nominal	3/8 0.375		7/16 0.4375		1/2 0.5	
Dowel pin	0.373	0.3745	0.4355	0.437	0.498	0.4995
Over/under	0.374	0.376	0.4365	0.4385	0.499	0.501

Today, these reamers are right up there on my most-used cutter list.

Unless you are as lucky as I was, it usually makes sense to buy reamers individually as the need arises (HSS only, of course). Beware of "expanding reamers." They seem to be mostly non-HSS, and in my experience they have never delivered a usable result.

5-38 HOLE BORING

The practical limit on hole cutting with Silver & Deming drills (Section 5-32) is about 1". Larger holes can be cut using a rotary table (see Chapter 8) or by using the method described here—a single-point bit in a boring head. The key advantages of single-point boring are *roundness* and *smoothness* of the hole and more precise hole *diameter* and *placement*.

Single-point boring is very slow, compared with other hole-making methods. It takes time and constant attention throughout. If you are enlarging, say, a small engine cylinder, concentricity is important, so the mill spindle has to be precisely aligned with the pre-existing bore. On the other hand, if you are boring a hole from scratch, start the job by drilling as large a pilot hole as possible—the less metal to be removed by the single-point tool, the better. Because the final placement of the finished hole will be determined by the boring tool, precise location of the roughed-out pilot hole is less critical than in the cylinder enlargement case.

5-39 BORING HEADS

The boring head shown in Figure 5-45 is a shop-made head, based on a vintage design by G. Thomas (Tee Publishing: *Model Engineer's Workshop Manual*). This took a lot of construction effort, but the end result works beautifully. Like most boring heads, it has dovetail slides and is positioned by a fine-pitch lead screw. It uses

FIGURE 5-45 Cutting a hole 1-1/4" in diameter. This started as a drilled 1" hole, here being enlarged to 1.25" in the second pass.

3/8"-shank carbide-tipped bits such as those in Figure 5-46.

One noteworthy feature of this design: The cutter shank is held in the shoe beneath the main body by a split circular clamp, which is rock solid, compared with the single set screw used in most commercial designs. The cutter bits can sometimes be used out of the box, but edge honing improves performance (also, for smaller-diameter holes, it may be necessary to grind a larger end-relief angle behind the cutting face).

FIGURE 5-46 Carbide-tipped cutter bits.

Commercial boring heads, such as the one shown in Figure 5-47, come in various diameters: 2", 3", 4", etc. For this type of head, the shank (R8 in this instance) is usually sold separately. The cutter bit is installed in one of two or three sockets on the underside and is secured by a set screw. In the example shown here, there is also provision for a transverse boring bar (installed crosswise) for larger holes.

If there are three set screws pressing on the dovetail gib strip, as in Figure 5-47, use only the middle one to lock the head (the other two are for routine gib adjustment—once set, leave alone).

Figure 5-48 shows a boring head with its cutter bit raised just clear of the workpiece. To increase the bore

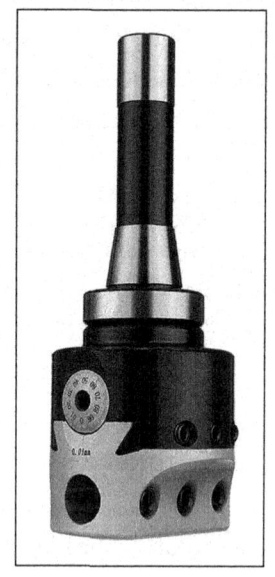

FIGURE 5-47 Boring head with R8 shank.

diameter on the next downward pass of the boring head, unlock the cutter slide, then offset it a few thousandths by turning the head leadscrew. Lock the slide in position, then run the spindle motor. Lower the boring head *very slowly* using the mill's fine downfeed control until the cutter is clear of the workpiece (through hole), or down to a depth stop if cutting a counter-bore, as in the diagram. Retract the spindle, then repeat as necessary.

In Figure 5-48, the face of the cutter bit is on the radius of the line between the hole centerline and the cutter tip—in other words, it has zero rake. In some cases, better results may be obtained with a rake angle of a few degrees, just as you will sometimes add some amount of back rake to an internal cutting tool on the lathe. Whatever the rake angle, it's import-ant to grind sufficient end relief to clear the bore—the smaller the bore, the greater the end-relief angle.

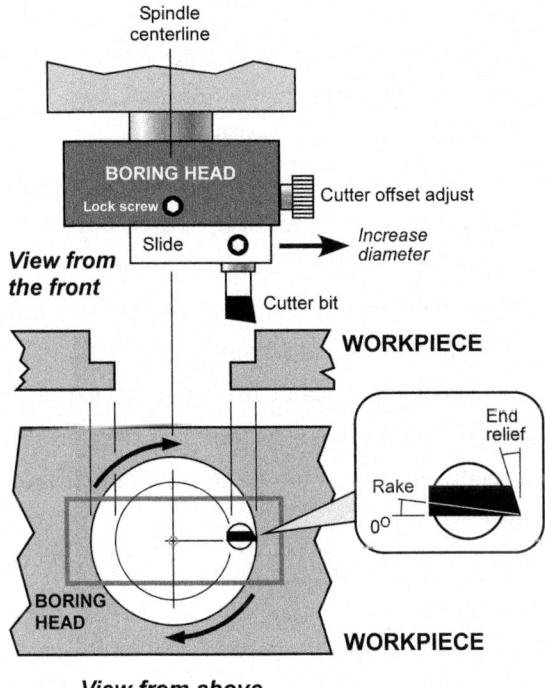

FIGURE 5-48 Boring head schematic.

Some cautions to bear in mind for all boring operations:

- Boring is unavoidably hazardous—a hard-to-see small cutter making wide sweeps.
- Always lock the boring head dovetails before running.
- Spindle speed is a matter of choice. I run boring heads at 150 to 200 rpm. This is a judgment call—faster speeds will amplify out-of-balance forces, but coffee-stirring speeds around 50 rpm can take forever to finish the job.
- Commercial boring heads are available with metric or inch lead screws—you need to ask.

5-40 MEASURING HOLE SIZE

If you have just drilled a hole, you know approximately what its diameter is—the nominal diameter of the drill plus a few thousandths depending on drill sharpness, chuck alignment, and hardness of the workpiece. A *reamed* hole has a more definite ID—exactly the nominal diameter of the reamer, plus maybe a two or three ten-thousandths.

Holes cut by an adjustable boring head are a different matter. You can measure them using the inside jaws of a caliper, but that's only good to within ± 0.002" or so. For more definite information, you need a hole gauge of some sort. Be wary of using drill shanks to check anything! The shanks are usually smaller than the nominal size of the drill. A better way is to machine a precisely fitting plug gauge from scrap material, and then check its OD with a micrometer, Section 5-41. For a quicker answer, there are two types of commercially available hole gauges: *small hole* (Figure 5-49) and *telescoping* (Figure 5-50).

A small-hole gauge typically has a hardened split ball, the OD of which is expanded by a tapered plug that is pulled into it by screw action.

The smallest of those shown in Figure 5-49 can be sized to fit any hole diameter from 0.125" to 0.2", the largest from 0.4" to 0.5". To use this type of gauge, place it in the hole, expand it to a snug fit, remove it, and then measure its diameter with calipers or (better) a micrometer. Repeat this two or three times to be sure of a reliable measurement.

The smallest of the telescoping gauges in Figure 5-50 covers the ID range from 5/16" to 1/2", the largest from 1-1/4" to 2-1/8". Care is needed to achieve reliable numbers with the telescoping gauge. Here's how:

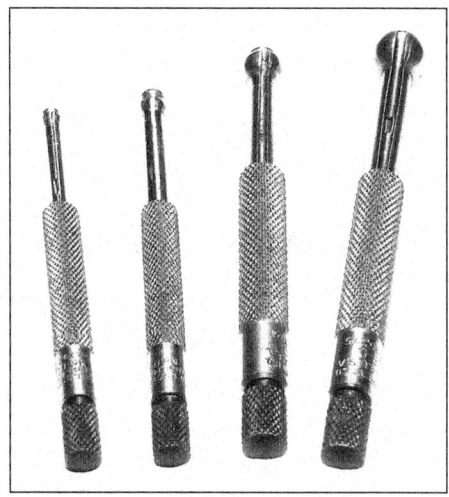

FIGURE 5-49 Small-hole gauges from 1/8" to 1/2" ID.

1. Release the rods by unscrewing the knurled cap at the lower end of the handle.
2. With thumb and forefinger compress the rods into the central casing; then hold them there by retightening the knurled cap.
3. Insert the gauge into the hole with the handle on the hole's centerline (or close to it).

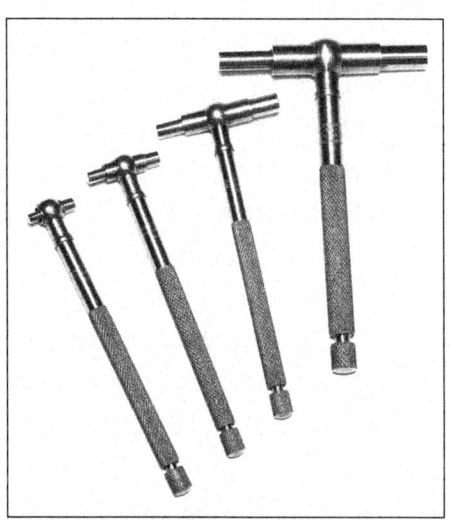

FIGURE 5-50 Telescoping hole gauges from 5/16" to 2-1/8" ID.

4. Release the rods; then wiggle the handle back and forth a few degrees to be sure the rods are properly in contact with the bore.

5. Retighten the knurled cap; then remove the gauge.

6. Measure across the rods with calipers or a micrometer.

It is a good idea to repeat the process a few times—you might be surprised by the spread of readings.

5-41 MAKE A PLUG GAUGE FOR BETTER PREDICTABILITY

Commercial shops usually have sets of precisely ground "pin gauges" from about 10 thousandths diameter up to an inch or so. Most small shops rely instead on the gauges shown in Figures 5-49 and 5-50, but there are times when greater accuracy is a must. The answer is to make your own *stepped plug gauge* on the lathe (Figure 5-51), using nothing more than a well-honed knife tool and a 0" to 1" micrometer.

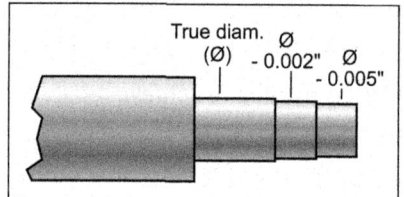

FIGURE 5-51   Shop-made plug gauge.

Starting with a scrap of steel about 1-1/2" long, turn the outer 3/4" length *exactly* to the desired hole diameter. Next, turn the outermost 3/8" down to minus 0.002"; then further reduce the first 3/16" of that length to 0.005" undersize. (These numbers are arbitrary and will vary, depending on the target hole diameter.)

The great thing about the multistep gauge is that it gives advance notice that you are nearing the right size, as compared with a single-diameter plug that (dismay!) goes into the bore and wiggles more than you'd like. If the first step, only, of your plug gauge enters the hole, you can safely open up by 0.002" and thereafter proceed by minute increments on the cross-slide.

5-42 TAPS

Aside from physical differences in flutes and chamfering (Figure 5-52), taps come in different pitch diameters designated H1, H2, and so on (often prefixed "G," meaning "ground threads"). The higher the number, the larger the diameter (for metric threads, "D" is used instead of "H"). What this means is that an H1-tapped hole might be an impossibly tight fit for a screw you just made using an adjustable button die (this may be no more than a temporary setback, provided you can tighten the die and rethread).

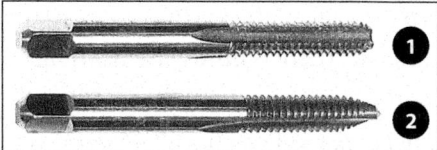

FIGURE 5-52 Taps. Aside from the obvious—nominal size and number of threads per inch—taps come with other visible differences in the number of flutes and lead-in chamfering. There are also nonobvious variations in pitch diameter. The examples here are (1) *bottoming* chamfer (with 2 chamfered threads), and (2) *plug* chamfer (with 4 chamfered threads).

Some suppliers do not say, or even know, what H number they have in stock. This is good to be aware of, but it's not usually a problem in small shop work. A middle-of-the-road choice is GH3. For taps, HSS should be a must, but carbon steel taps are still available, especially in sets. Don't be tempted—economy taps, carbon or even HSS, are likely a false economy.

The two most useful tap styles are:

- **Bottoming chamfer.** Taper ground at the leading end, just one or two threads. Better for tapping blind holes, not so good as a plug chamfer for starting the thread (but you can usually get by with just a bottoming tap if you countersink the hole a little using a center drill).
- **Plug chamfer.** This type of chamfer has three to five of the leading threads ground in a gentle taper. A good general-purpose tap for most applications, except for tapping blind holes. You will rarely, if ever, need a so-called taper-chamfered tap (one with the first seven threads chamfered).

5-43 HOLE SIZES FOR TAPPING

There is nothing sacred about the hole sizes recommended in the usual thread tables. They are designed to give you a 75% depth of thread, but 75 isn't a magic number. According to industry sources such as Greenfield Tap and Die, there is little gain in strength over 60% depth. This means you can drill one or two number sizes larger than the table. *Result: Less chance of tap breakage*, less effort, less stress on the workpiece.

5-44 TAPPING OPERATIONS

When threading a drilled hole, the tap has to be *perfectly aligned* with the bore. If you are positioning the workpiece by dead reckoning (counting dial graduations), the usual practice is to drill and tap as a combined oper-ation—you don't move on to drill hole 2 until you have tapped hole 1. It is more convenient if the mill has a digital readout, in which case you would drill all the holes in one pass, then go back to the same hole positions with a tap instead of the drill—no need to swap tools every time.

- **Tapping Method 1.** Small taps up to, say, M6 or 1/4" can be held in the drill chuck, which can then be turned by the left hand at the same time as the quill is eased downward with the right (this ensures proper seating of the tap and elimi-nates upward pull by the return spring). The aim here is to feed the tap in three or four turns, *absolutely square*, follow-ing which the tapping can be completed using a regular hand wrench (Figure 5-53). Note, though, that there are two problems with this method: (1) The chuck may not grip tightly enough to avoid slippage. (2) If yours is a keyless chuck, the tap-ping rotation (clockwise

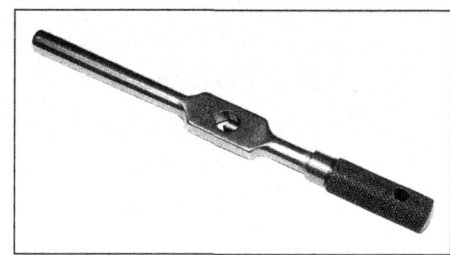

FIGURE 5-53   Starrett tap wrench.

looking down) tends to loosen the chuck. That said, it can be a quick way to get the job done without a special holder.

- **Tapping Method 2.**
  This is an adaptation of
  Method 1. Instead of
  relying on the gripping
  power of your keyless
  chuck, turn the mill
  spindle itself by using
  the *drawbar wrench*. As
  before, you still need to
  maintain a light down-
  ward pressure on the
  quill to keep the tap

FIGURE 5-54 Rotating the spindle with a ratcheting wrench.

  properly engaged. For a more convenient means of turning
  the spindle, consider using a ratcheting wrench, as Figure
  5-54. In this example, the wrench is 21 mm.

- **Tapping Method 3.** Larger taps with *center-drilled shanks* can
  be kept in alignment by
  a shop-made pointed
  steel rod in the chuck
  (Figure 5-55). Engage
  the pointed rod by
  downward pressure on
  the quill; lock the quill,
  make a half-turn or
  so of the wrench, then
  unlock and lower the
  quill again. Repeat as
  necessary.

FIGURE 5-55 M8 threading with a center-drilled tap.

- **Tapping Method 4.**
  There are (or there used to be) tap holders such as the one
  shown in Figure 5-56. Basically, this is a conventional tap
  holder with a short cylindrical shank that slides in a sleeve.

The smaller-diameter end of the sleeve can be held in either a drill chuck or a collet. The basic idea is that it allows you to feed the tap several turns into the work without needing the quill to be lowered. You can then finish the job using a regular hand wrench, as described above.

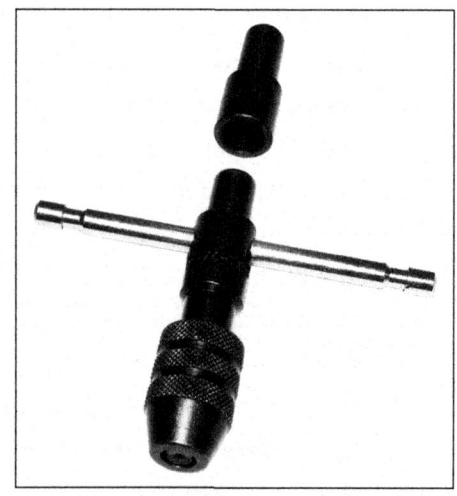

FIGURE 5-56 Sliding tap holder for the chuck.

In some cases, tapping can be done under power, but this is not something to try without a lot of practice on the mill. For one thing, reversing is *not instantaneous*, so be careful tapping blind holes. For another, be sure the workpiece is reliably secured to the vise or table, so that it *cannot be pulled upwards by* the tapping action. Also, be sure the quill locking lever is free, and—important—consider easing the quill down in sync with the threading action (otherwise, the quill return spring can overwhelm the hold of the tap in the workpiece).

Even for hand tapping, consider using a *tapping fluid*. Any cutting oil is better than none, but most users find Castrol's Moly Dee the most reliable for threading steel.

CHAPTER 6

More Fundamentals

CONTENTS AT A GLANCE

6-1 CONVENTIONAL MILLING VERSUS CLIMB MILLING

This is discussed in Section 6-4, but it's important enough to emphasize here as we begin this chapter. We take for granted that milling cutters will rotate clockwise when viewed from above. For *conventional* (or *up*) milling, we would expect the workpiece to move in the opposite direction to the cutter teeth at the point of contact. But here's the tricky part: The "right" direction depends on which surface you are milling—front face versus back face (X axis) or left face versus right face (Y axis).

The "other" way of milling, known as *climb* (or *down*) milling, has the work moving in the same direction as the cutter teeth at the point of contact. Plus factors of climb milling are less heat in the workpiece, lower cutting force, and better surface finish (chips fall behind the tool). On the negative side—*caution*—the cutter can move the table *unexpectedly* (by taking up backlash) and that it allows only light cuts.

6-2 REALLY NEW TO ALL THIS?

If so, the following should be helpful. Some of it has been said elsewhere, but repetition is no bad thing. See Chapter 3 for words on materials choices, cutting fluids, and other basics.

6-3 MILLING FEEDS AND SPEEDS

What spindle speed should I use? What cutter size? What depth of cut? How fast to move the table? In production operations, all these things are well known—they are optimized for the best compromise between throughput and tool life. In the model shop, though, we have a quite different situation. For one thing, most every work session is different from the one before. For another, we don't usually have reliable knowledge of cutter quality and workpiece material. Finally, looking at weight, rigidity, and speed range, we can see that there's no real comparison between a 1-hp model shop mill and its 10-hp industrial counterpart.

Spindle speed is the key factor. Cutter manufacturers will recommend a "cutting speed," in feet per minute (SFPM), but that's not directly helpful (cutting speed means speed at the *periphery* of the tool, which depends on both spindle speed and cutter diameter). Therefore, the best we can do instead is work from a simplified table of spindle speeds, like those in Table 6-1, for various materials and cutter sizes. Bear in mind that two frequent source of problems is small cutters running too slow, and large cutters too fast.

TABLE 6-1 Suggested spindle speeds for HSS mills

Material	Cutting Speed (ft/min) SFPM	Cutter Diameter and Spindle Speeds			
		1/4"	3/8"	1/2"	3/4"
O1 tool steel	50	750	500	400	250
Cast iron	60	900	600	450	300
W1 tool steel	85	1300	850	650	400
Mild steel*	100	1500	1000	750	500
Brass/aluminum	250	3750	2500	1900	1250

These are only starting figures—workpiece hardness and cutter quality/sharpness vary widely. Most mills do not offer these exact speeds, especially the highest speeds suggested for brass and aluminum—use the nearest available. Carbide mills should be run faster, between X2 and X4.
*Mild steel means any low-carbon steel (1018) and leaded alloys (12L14).

6-4 WHICH WAY TO MOVE THE TABLE?

This depends on two factors: (1) Which face of the workpiece is being machined—front, back, left, or right, and (2) whether you are milling *conventionally* or are *climb* milling.

In conventional milling (aka up milling), the workpiece is fed against the cutter's rotation (Figure 6-1). Climb milling (down milling) has the workpiece going in the same direction as the cutter. Unless drag is applied to the table (light clamping), the cutter *can itself apply traction* to the workpiece, causing unexpected "self-feeding" motion due to backlash in the table drive. In the worst case—very sudden and extreme—it can cause the cutter to *shed teeth*. That said, climb milling causes less wear on the tool and can sometimes deliver a better surface finish. So *think before climb milling—proceed with care!*

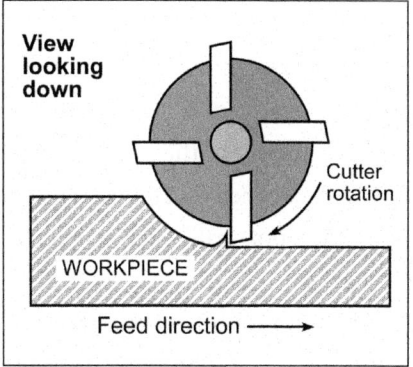

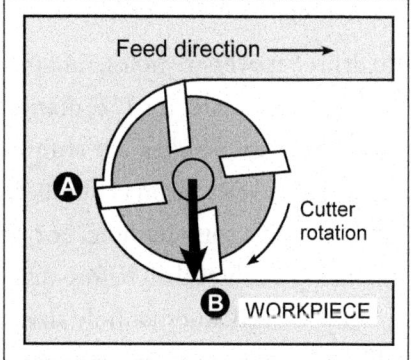

FIGURE 6-1 Conventional milling direction, viewed from above. In this illustration, the cutter is machining the back surface of the workpiece, so the table moves to the right. For the front surface, the opposite applies. The same logic applies to front-to-back motion of the table when machining left- and right-hand surfaces—just make sure the feed direction opposes cutter rotation on the side in question.

FIGURE 6-2 Why 4-flute cutters cut slots oversize. At the moment in time shown here, tooth A is doing most of the work. The reaction to the cutting force at A is in the direction of the arrow. This pushes the cutter down, causing tooth B to dig into the side wall—hence the oversize cut.

6-5 SLOT CUTTING

Don't cut slots with a 4-flute cutter! Why? Because they cut oversize (Figure 6-2) and leave the sides of the slot rough. If the slot width is to be, say, 5/16", use a reground (smaller) 5/16" to precut the slot; then make two finishing passes, one to the left, one to the right. This problem doesn't usually arise with 2-flute cutters, which is one reason why they're sometimes called "slot drills."

6-6 DEPTH OF CUT

Assuming the spindle speed is correct for the material and cutter diameter (see Table 6-1), depth of cut is usually arrived at by experiment. Always start on the light side, even as little as 0.02", with a slow feed rate on the table. Increase both cut and feed in small increments, stopping at the point where there is noticeable vibration or other undesirable effects.

6-7 DRILLING AND REAMING

All drills cut oversize holes, ranging from about 0.002" for a 1/16"-diameter drill to about 0.005" for 1/2" diameter. This applies even to new drills.

Whenever possible, use stub-length drills instead of the standard jobber length. They stay on center better and, because they don't flex so much radially, drill a rounder hole. For precise work, "spot-drill" each hole location using a center drill before drilling to size.

If you need an exact hole size, drill a few thousandths undersize; then follow with a reamer (or, if the hole is large enough, bore to size). Reamers are expensive, so consider buying individual sizes as needed instead of sets. Do not be tempted by adjustable reamers. They sound good—but, like broken-tap extractors, have never been known to work well. Run reamers about one-quarter the speed of the corresponding drill size, using plenty of cutting oil. Never run them in reverse.

If a reamer isn't available, drill a few thousandths undersize as before; then redrill with a brand-new drill of the nominal hole size, using very little downward pressure.

6-8 THREAD TAPPING

When threading a drilled hole, it is essential to align the threading tap properly in the bore. The mill is often used for this purpose, ideally with a dedicated tap holder or, for production work, an auto-reverse tapping attachment. Typically, all you need to do for sizes up to M6 or 1/4" is to *swap the drill for the tap*, then *hand-turn the chuck* while keeping a gentle downward pressure on the quill. After two or three turns, the tap may be stable enough for completion using a hand wrench. Tapping can also be done under power, which is sometimes kinder on the tap than turning by hand—but this is a judgment call based on experience. For either method, consider using a *tapping fluid*, depending on the workpiece material. It's worth repeating that Castrol's Moly Dee is the most reliable for threading steel.

When power-tapping, bear in mind that the spindle doesn't reverse instantaneously, so be careful tapping blind holes. Be sure the quill locking lever is free, and start trial work with the lowest spindle speed.

6-9 PVC IS A GREAT SUBSTITUTE FOR METAL

"Type 1" rigid PVC is far less expensive than any metal and is available in accurately dimensioned sheets, bars, rods, and discs. I have used PVC on many projects over the years, not only for templates, but also for practice-machining to check out the process and dimensions before cutting metal. Aside from solid-modeling CAD and 3D printing, this is the fastest way I know to produce a realistic "3F" model (*fit*, *form*, and *function*).

Here are a few approximate 2022 prices for PVC (expect to pay four times as much for aluminum):

12" x 12" x 1/8"	$7
6" x 6" x 1/2"	$7
1" x 1" square	$7/ft
2" diameter	$13/ft

An excellent supplier of PVC is McMaster Carr. The company has every conceivable size and shape in stock and in most cases can get it to you overnight if you are near one of its locations.

Key facts about PVC:

- It machines beautifully at high spindle speeds, with no significant cutter wear (but watch out for overheating).
- Blanks of the material can be rough-cut to size in seconds using a table saw or chop saw.
- If you need a block of PVC, but all you have in the bin are thin sheets, stack them using plumbing adhesive.
- *One caveat:* When heated, PVC expands four times more than steel—and it cools more slowly, so take care when checking measurements.

- Holes drilled in PVC tend to shrink as the material cools. Sometimes this can be a problem, but not when you need to hold locating pins firmly in place.

6-10 CUTTING AT EXACT ANGLES

Sooner or later, a project comes along that calls for something other than right-angled sides. Setting up the mill to machine "non-90" angles with respectable accuracy always requires extra planning and care, and sometimes imaginative workarounds. One off-the-shelf solution is a standard set of angle blocks (Figure 6-3).

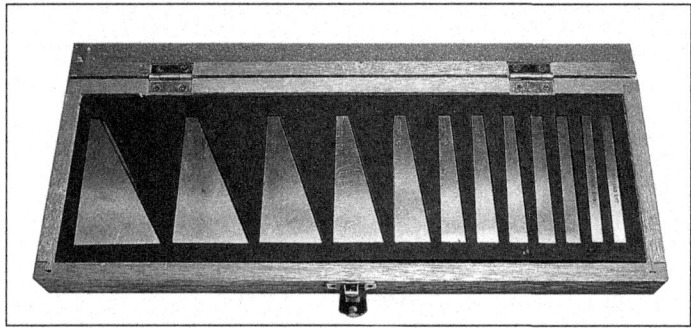

FIGURE 6-3   Angle block set. This is a standard set of twelve 3"-long angle blocks ranging from 1/4° to 30°. By stacking these 3"-long blocks together, you can assemble any angle from 0.25° to 115.75° in 0.25° increments.

Figure 6-4 shows a machining project calling for a ramp AA angled at 32-1/4° to the baseline BB. The first step might be to scribe and cut off the surplus metal, probably by bandsawing, following which one of the edges of the workpiece would be machined flat. The machined edge of the workpiece would then be set on a (pre-assembled) 32-1/4° stack of angle blocks.

In some situations, this might be a workable arrangement, but it can be a three-hands job trying to keep the workpiece and angle stack together, while at the same time tightening the vise. There is a better way, described in Section 6-11.

Practical problems with angle blocks are these:

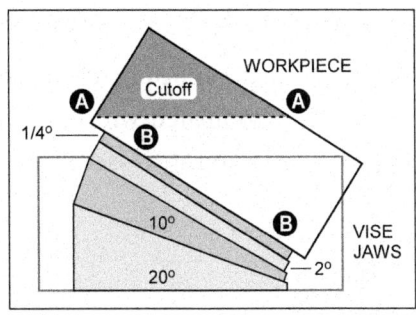

FIGURE 6-4 A 32-1/4° angle block assembly.

1. The stack tends to fall apart just when you have the setup in the vise ready to be clamped up (a grease film between the blocks can help).

2. The blocks are 1/4" thick and therefore difficult to use on thin-sheet material (not impossible, but you will need additional clamping above the angle stack).

3. The stack always pushes the workpiece further out of the vise than you would like for solid clamping.

4. The smallest increment is 1/4°, often adequate, but sometimes the job calls for greater precision.

All of the above are the reasons I have rarely used my inherited set of angle blocks—maybe 10 times in the past 15 years, for inspecting angles, not setting up for machining. If you don't yet have a set of angle blocks (around $150, at 2022 prices), my suggestion is to leave it on the maybe-sometime list of tools.

There is any easier, sometimes better, way to do the job shown in Figure 6-4: Make your own angle template from scrap material, *thinner than the workpiece* for easy setup.

If you have a DRO and suitable material, this is literally a 20-minute job. Even if you are moving the table by dead reckoning from the dials (no DRO), it won't take much longer—but use "one-way" thinking to avoid backlash errors. But first, think about PVC (Section 6-9). PVC is an excellent material for templates—and numerous other tasks around the shop.

6-11 MAKING A 32-1/4° TEMPLATE

The template described here, and shown in Figure 6-5, was made of 1/8"-thick PVC. Compared with the angle block assembly, it allows a much improved gripping area—see Figure 6-6.

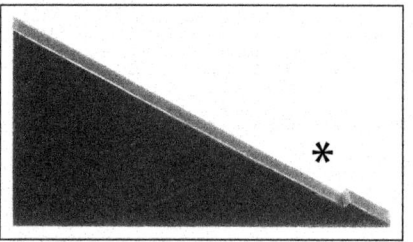

FIGURE 6-5 PVC angle template. The notch (optional) indicated by the asterisk helps ensure that the bottom edge of the workpiece sits firmly on the template.

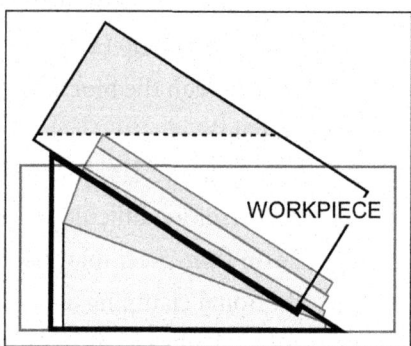

FIGURE 6-6 Larger gripping area with a shop-made PVC template.

The dimensions given in Figure 6-7 are arbitrary, chosen to make the template small enough to fit comfortably in a 4" vise. When you have chosen the X axis spacing of the two locating holes, 2.2" in this example, multiply that value by the tangent of the desired angle, 32-1/4°.

$$Y = 2.2 \times \tan 32.25 = 2.2 \times 0.631 = 1.388$$

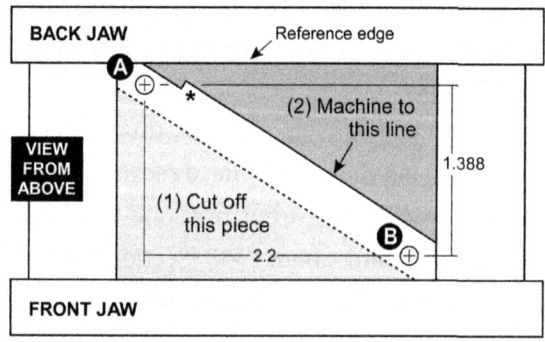

FIGURE 6-7 Template dimensions.

To make the template, rough-cut a rectangle of 1/8" material; then set it in the vise with one of its longer edges uppermost. Trim the exposed edge with an end mill to form a smooth "reference edge." Flip the rectangle upside down; then trim the other long edge (this will allow it to be held firmly in the vise when laid flat, for the next operation).

With the PVC held flat in the vise, on parallels, drill locating holes A and B for the stop pins (Figure 6-8). If you follow the suggested dimensions, the pins will set the workpiece in the vise at exactly 32-1/4°. The holes here were drilled 1/8". This, in PVC, gives a snug fit for 1/8"-diameter locating pins—no need for reaming if the drill is in good condition. Hole diameter is arbitrary, depending on any thin rods you have on hand, such as drill rod or even wire from a coat hanger.

FIGURE 6-8 Drilling the locating holes.

Cut off the surplus triangle (item 1 in Figure 6-7); then insert the two locating pins. Clamp the template firmly in the vise with the pins sitting on the vise jaws or, optionally, held off by about 1/8", as in Figure 6-9. Machine the exposed edge, not forgetting the notch at the end. If you take the pins out before milling, you will be able to lower the end mill, thus reducing the height of the template. This might be helpful if you are using a smaller-than-usual vise.

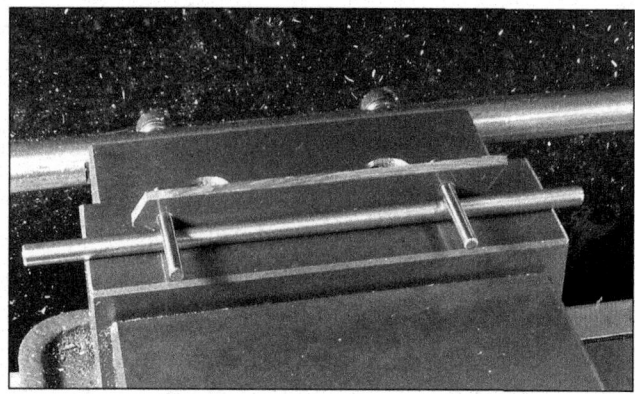

FIGURE 6-9 Template ready for end milling. In this example, the pins are raised above the jaws by a 6"-length of 1/8" drill rod.

6-12 ANOTHER ANGLED-CUT EXAMPLE

Figure 6-10 calls for a 15.3° ramp at the end of a 2-1/2" block, which is too long to be held reliably on a single template such as the one shown earlier in Figure 6-5. One way to deal with this would be to make two identical templates, one for the back jaw, one for the front. Another way is to use a *swivel base* under the vise to set the ramp angle. For this alternative method, the block would be clamped in the vise with its long dimension vertical. The question now is, *How can a precise 15.3° rota-*

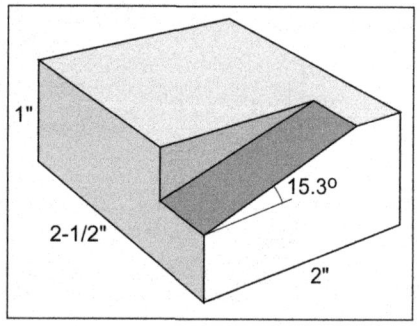

FIGURE 6-10 Consider using a swivel base for this angle cut.

tion of the vise be achieved? We need better than a ± 1/4° estimate, which is probably the best you can do with graduations on the base.

In the following, it is assumed that the block to be machined has been trued on all sides—flat and 90° all around (see Section 6-14).

The first job is to make a template from scrap material, possibly PVC, along the same lines as in the preceding sections. The dimensions for this template are given in Figure 6-11. This template is easier to make than the one in Figure 6-5, because there is no need to remove the angled edge.

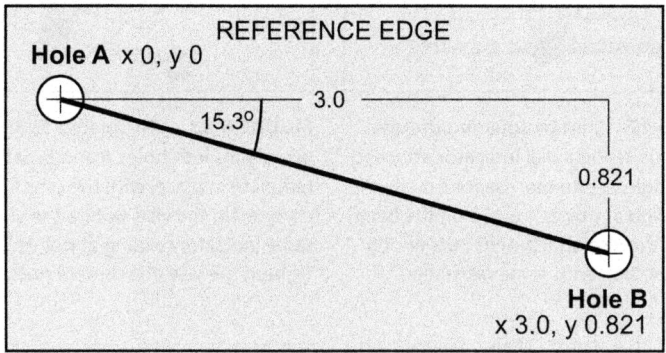

FIGURE 6-11   A 15.3° template.

All you need here is a parallel-sided scrap of material, with one of its edges (the reference edge) flat and true for reliable indicating.

Fasten the swivel base to the table. Fully tighten its attachment nuts (alignment is not important). Install the vise on the swivel base, squaring it by eye; then snug its attachment nuts. Indicate the back jaw, as shown in Figure 6-12, tapping the vise into exact alignment.

Prepare for the next step by inserting dowel pins into holes A and B of the template, ensuring that the pins are secure. With the dial indicator against the template's reference edge, tap the vise around for exact alignment (Figure 6-13). Fully tighten the vise attachment nuts.

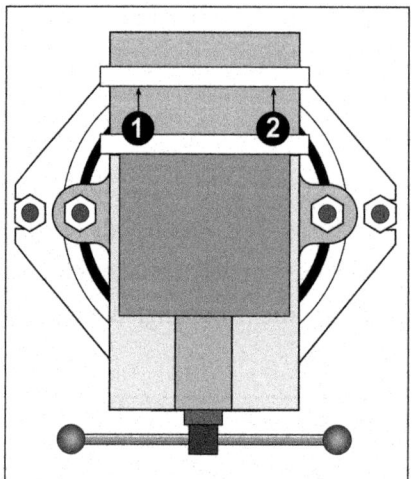

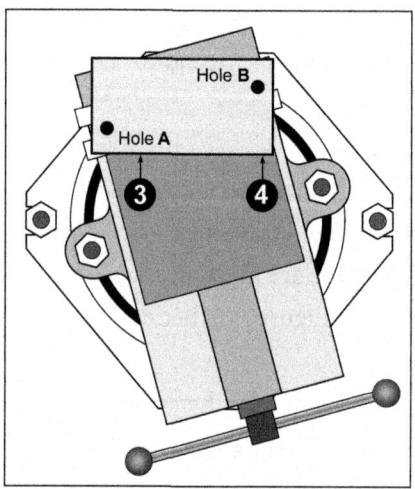

FIGURE 6-12 Start by squaring the vise to the X axis. Using a dial indicator attached to the spindle, rotate the vise for exactly the same reading at points 1 and 2 on the back jaw. Snug the vise attachment nuts enough to allow rotation with some resistance.

FIGURE 6-13 Vise rotated 15.3°. Insert dowel pins into holes A and B. Set the template in place, with the pins firmly clamped in the vise. Rotate the vise for the same indicator reading at points 3 and 4. Tighten the vise attachment nuts.

For the final step, install the workpiece in the vise; then machine the ramp feature (Figure 6-14).

6-13 BACK TO BASICS

Milling is basically the process of removing material from a block, a sheet, or a bar of metal. In most cases, the workpiece starts out in rectangular form, with at least its reference faces nicely squared with respect to each other. That's certainly how it can be if you are shopping for a specific project, and your metal supplier will cut off an inch

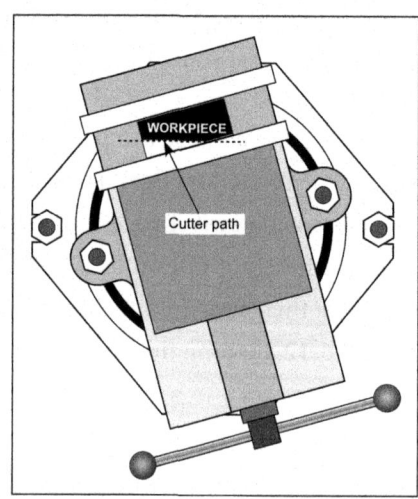

FIGURE 6-14 Remove the template, and set up the workpiece. Clamp the workpiece in the vise; then end-mill the ramp to the desired depth.

or two of this or that bar stock. Most of us, though, start with whatever material is on hand, in some cases tweaking designs to avoid going out for additional supplies. Inevitably, this calls for additional prep work, first cutting the stock to size on the bandsaw, then machining it to a suitable shape from which to proceed.

6-14 START BY SQUARING THE MATERIAL

To say that there's a big difference between two- and three-dimensional workpieces may be stating the obvious, but here goes: Workpieces that are cut from sheet material can be called 2D. While they might be carefully sized, and linear where it matters, we usually don't pay much attention to the edges themselves—a little roughness here and there may be a secondary concern. Squaring sheet material has been considered elsewhere (Section 4-5).

The workpiece becomes "more 3D than 2D" as its thickness increases, resembling a block rather than a sheet. Squaring a block calls for a quite different approach, often the beginning stage of many projects. While it may not always be necessary to square all six sides of a block before machining other features, it can sometimes be the least frustrating way. Even if you are starting a project by squaring just two or three sides, it may be helpful to understand the entire "all-square" process.

Imagine a workpiece that starts out as a nominally rectangular block 2" x 1" x 1", rough on all faces, probably cut by bandsaw. Squaring the material is a seven-step process:

- **Step 1.** To machine the first face, you will need a "pusher" between the moving jaw and the workpiece. This can be anything that will not apply a steering force of its own.

  The top photo shows a hemispherical pusher turned from, say, a 3/4"-diameter aluminum rod. Another pusher option for step 1 could be a length of semi-rigid 1/4" plastic rod, such as Delrin—anything that will press the rough block firmly against the back jaw of the vise, the reference plane, shown in the diagram. If the workpiece is below the vise jaws, raise it using a single parallel under the back edge, as in the lower photo.

STEP 1

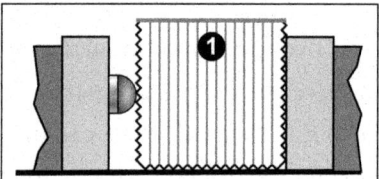

Parallel (if needed)

Machine the top surface with an end mill or fly cutter. To reduce the number of passes, use the largest-diameter cutter you have on hand.

THE WARP FACTOR

Most metals change size and shape when machined. In many cases, the change is scarcely noticeable, but that's not so with cold-worked steel, such as 1018, the standard offering from most stockists (bright finish, oily surface). This is not usually a showstopper if you plan for it—just be aware that flat bars may not stay flat. Skimming only a few thousandths off the outer surface relieves stresses "baked in" when the material was drawn, rolled, or otherwise pounded into shape. Hot-worked steel, such as A36, may be better in that respect, but don't bank on it.

- **Step 2.** Turn the block upside down; then *tap it down* onto parallels just below the top surface of the vise jaws—enough gripping surface to hold the block firmly, but not so much that the block is skewed off the parallels. Tapping down is important (use a soft metal mallet): Both parallels need to be firmly in place, enough to resist being pushed easily from side to side. *Machine the top surface.*

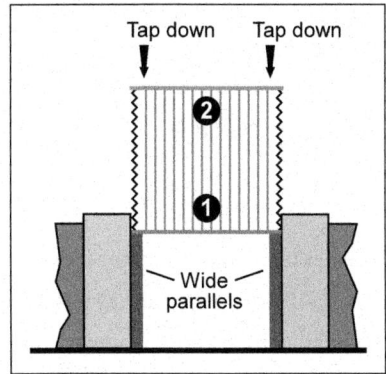

STEP 2

- **Step 3.** Turn the block 90°; then set it down in the vise as deep as possible. Raise it with parallels if necessary. *Machine the top surface.*

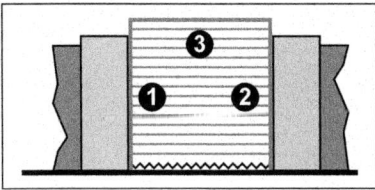

STEP 3

- **Step 4.** Turn the block upside down; then *tap it down* onto parallels. *Machine the top surface.*

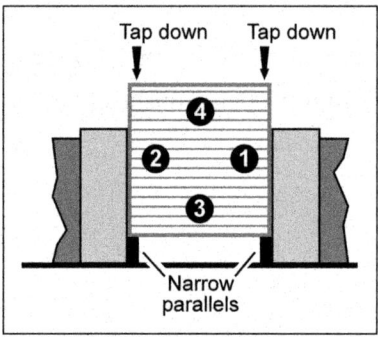

STEP 4

Important note: Once the fourth face has been machined, the *cross section through the block is square at all four corners*. Conventionally, at this stage the block would be set lengthwise on parallels, and the ends would be machined square by side-cutting with an end mill. The following procedure uses no side cutting, and it takes a little extra time for one additional pass. However, it has this key advantage: It delivers the *same surface finish* on the ends as on the faces (which is not true of side cutting). Additionally, if you were using a fly cutter (or other surfacing tool) up to this point, you can finish the job *without changing tools.*

- **Step 5.** Set the block with its long edges vertical. There is no need to be overconcerned about precise "verticality"—just set it as square as you can by eye. To be sure of its orientation in the following steps, checkmark the front face. In both the diagram and the photo, it is deliberately out of square to show how this is corrected in the next steps. *Machine the top surface.*

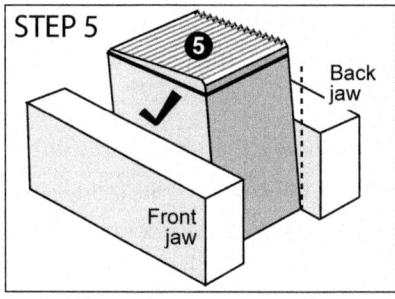

- **Step 6.** Flip the block upside down; then rotate it so that the checkmark can be seen at the right side of the vise. (*Note: If in step 5 the block had been tilted to the right, the check mark would be seen on the left side.*)

    Tap the block down onto a single parallel against the back jaw of the vise. (The photo shows the block well clear of the front parallel, due to the leftward tilt in step 5.) *Machine the top surface.*

Once the sixth face has been machined, all faces except #5 are square to each other. This is easy to correct, but it does call for an extra step, Step 7.

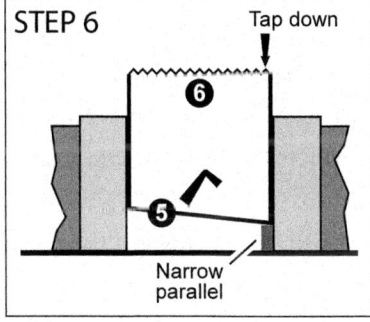

Why step 7? Why not set the block on parallels after step 4 and side-cut the ends? True, you could do this, but side cutting gives a very different surface finish.

- **Step 7.** Flip the block upside down; then tap both front and back edges down onto narrow parallels. *Remachine face #5 to correct uncertainty about tilt in step 5.* All six faces are now square.

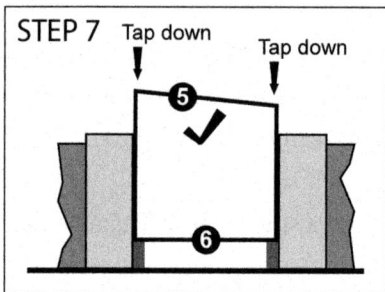

6-15 WORKING ON BAR STOCK

To most machinists, this suggests "round work," meaning jobs for the lathe. Once in a while, though, we use the milling machine to cut gears and splines, mill hex flats and square ends on rods, cut keyways, drill through-holes, etc. Work holding is the main issue. Section 5-19 shows one of the traditional ways of holding cylindrical work using a V-block with an angled clamp screw. More conventionally, you might hold bar stock between a pair of opposing V-blocks in the vise. The problem with V-blocks is that the machining action—slot, hole, whatever—is not on a *defined radius* of the cylinder, so there's no obvious way to relate what you just did to any subsequent work.

For an easy workaround, consider the *block chuck*, a special holder for 5C collets ranging from 1/16" to 1-1/8" capacity, Figure 6-15. The blocks measure 2-3/4" long, 1-3/4" across the flats. (**Note:** 5C collets are available for round, square, and hexagon cross-sections.)

Block chucks usually come in *square* and *hexagonal* cross sections, often sold in pairs, sometimes with a "closing lever," basically a cam action that grips the work firmly without the need for a special wrench. This sounds convenient, and it usually is, but it limits the "held" length of bar to about 3-3/4", the length of the block plus about 1" within the lever body.

FIGURE 6-15 5C block chucks.

If the workpiece protrudes more than 1" beyond the back of the collet, you will have to use a closing ring instead of the lever. This needs a special pin wrench, probably not supplied, or you will have to tighten the ring by "tapping it round" in the time-honored way with a soft metal peg and hammer.

A word on sources: The blocks sell for $150 or more in the major catalogs, but almost identical items are on eBay and Amazon for less than $50. Also, don't be tempted to buy an entire set of 5C collets, unless you would also use them on your lathe. Individual sizes are available for less than $15 each.

FIGURE 6-16 Cutting a keyway in 3/4" steel, using a 5C collet in a square block chuck.

FIGURE 6-17 Cutting a hex head on a 3/8" screw, using a 5C collet in a hexagon block chuck.

6-16 THE SPIN INDEXER

Cutting a gear is a lot more demanding than milling hex heads and flats: It calls for some means of stepping a disk blank around in specific angular increments defined by the gear's "tooth count." Aside from CNC, two ways to do this are a rotary table, Chapter 8, and the less-costly device discussed here, the *spin indexer* (unsurprisingly, often shortened to *spindexer*). Spin indexers such as the one shown in Figure 6-18 are mostly made for 5C collets (but they are now also available for ER collets). Indexers for 5C collets like this go for as little as $50.

Indexing pin

FIGURE 6-18 Spin indexer with 5C collet.

The spin indexer has been around for years, originally intended to be held by a magnetic chuck on a surface grinder. I no longer have a surface grinder, but I still have the indexer. Due to its grinder heritage, the snag with this device is that only its face and underside are machined true to the collet tube. Everything else is painted casting, and there are no holes for T-bolts. Tramming the indexer every time you want to use it is enough of a deterrent to come up with a better solution, such as holding it in the milling vise, as in Figure 6-19. But for that to work well, the sides of the indexer casting *have to be parallel*, exactly at right angles to the face. Machining my indexer for that purpose took no time at all, because I could hold it in a larger milling vise, with its face against the back jaw.

FIGURE 6-19 Modified spin indexer with machined edges.

Instead of a vise, if yours is a smaller machine, consider using a slotted angle plate, Figure 6-20. Secure the angle plate to the table, tramming it carefully to be sure its vertical face is precisely aligned with X axis (left-right motion). Secure the indexer to the angle plate using a couple of screws, e.g., 3/8" x 6". Check the machined flat surface of the indexer using a dial indicator held in the chuck, aiming for ±0.002" or better when traversing from side to side. Fully tighten the screws, then skim the indexer edges as in Figure 6-21. The indexer will now be ready for installation in the milling vise, with reasonable certainty that it will need no further tramming.

As an alternative to the vise, consider clamping the indexer to the table using standard clamping kit hardware, Figure 4-3. The downside of this is the need for time-consuming alignment of indexer and table. This problem goes away if you machine key slots in the indexer base, just like the key slots that came with the vise, Figure 4-10. Cut the key slots using the Figure 6-21 setup. With keys in place, you can quickly clamp the indexer to the table, confident that the collet tube is properly aligned.

FIGURE 6-21 Machining the indexer edges.

FIGURE 6-20 Indicating the bottom surface of the indexer. The aluminum block was made to fit in the machined rim of the bore.

6-17 USING THE SPIN INDEXER

The indexer dial is locked in position by removable pin through the front face, indicated by the arrow in Figure 6-22. The dial has 36 holes, which suggests nothing finer than 10 steps. In practice, however, the indexer's vernier-style system allows you to step around in 1° increments. This gives 10 or more "tooth counts" for gear cutting: Tooth count (T) x step angle = 360, where both T and the step angle are whole numbers, e.g., 120 x 3°, 90 x 4°, 72 x 5°, 60 x 6 . . . 30 x 12°, 24 x 15°. Values of T not in this series call for even finer angular steps, maybe using a rotary table, Chapter 8.

FIGURE 6-22 Vernier for 1° indexing. Here, the dial has been rotated to align 7 (actually 70) with the Vee datum between numerals 4 and 5 on the indexer body. We know that the dial is at exactly 70° because the 0 hole in the casting (see the arrow) is aligned with a hole in the dial. Only at this point can the pin go through the dial plate, preventing further movement. If the pin were to be removed, and the dial turned a tiny amount counter-clockwise, the pin would go through hole 1—that's 71°—but not through any other hole. At each setting, turn the knurled knob to lock the collet tube.

6-18 A PRACTICE WORKPIECE

For a low-risk, low-cost trial piece to get experience with your new machine, you might start by squaring a 1-1/4" cube of aluminum or PVC, as described earlier; then machine it as suggested in Figure 6-23. The end result has no practical function, but it does introduce operations you'll find useful in everyday machining, such as milling at an exact angle, slot cutting, drilling, tapping, and counterboring. The procedure suggested here assumes you don't have a DRO, so you will be positioning the table by counting dial divisions, described in Section 3-6 (100 divisions = 0.1"). The order of events is not set in stone, but it's possible that too much of a departure will paint you into a corner. If nothing else, this exercise will show how a DRO can be a huge bonus in time-saving and accuracy—no need to worry about backlash; just look at the numbers.

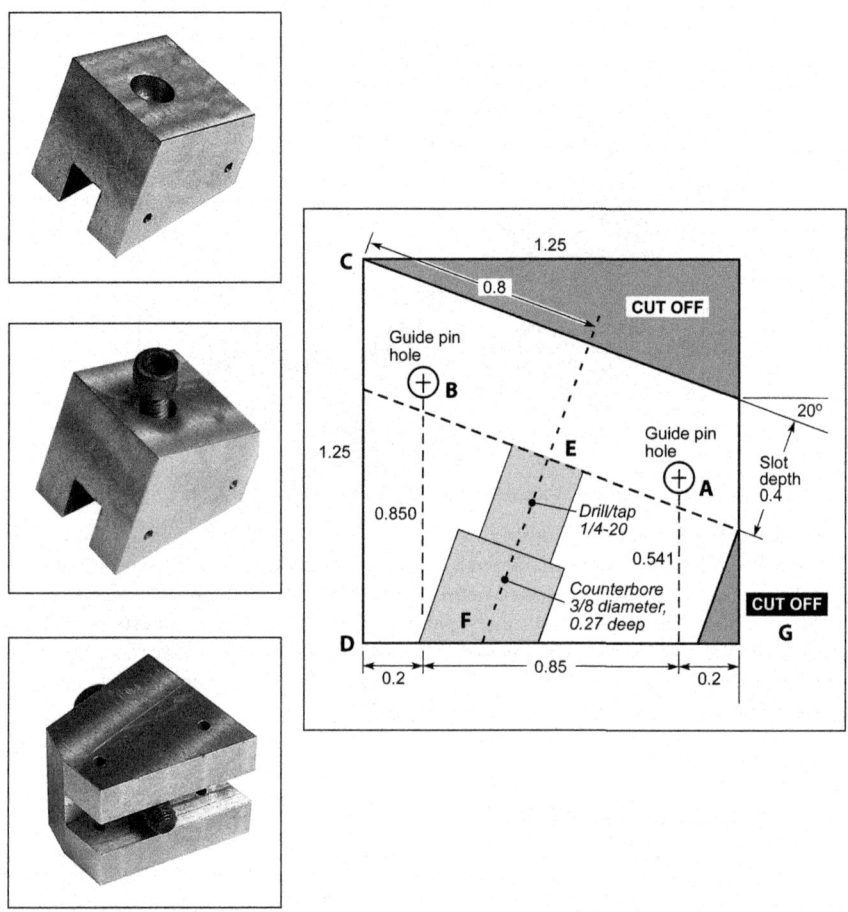

FIGURE 6-23 Practice piece. Dimensions in inches. The three photos to the left show views of the finished project.

You will need the usual metal shop hand tools, plus:

- 3/8"-diameter 2-flute (center-cutting) end mill for slot cutting. For initial squaring the block, you will probably use a fly cutter or a larger-diameter end mill. (A 3/8" mill could be used instead, but that would take many passes.)
- 3/8" collet or R8 end mill holder.

- Edge finder (a 0.2"-diameter tip is assumed); see Section 5-11.
- Vernier or digital caliper (nothing fancy; think $20 or so).
- Two metal guide pins, about 2" long, diameter a little less than 1/8".
- Twist drills (hardware store sizes will do for this project), 1/8" and 13/64".
- #2 center drill (Section 5-26).
- #1/4-20 tap, standard plug chamfer (Section 5-42).
- Any 1/4-20 socket head screw.

6-19 DRILLING THE GUIDE PIN HOLES

Follow these steps:

1. As shown in Figure 6-24 (1) and also Figure 6-25. With the edge finder in a chuck or collet, and with the spindle running at 500+ rpm, move the table *right* until the edge finder is tripped by contacting the block. Note the dial reading.

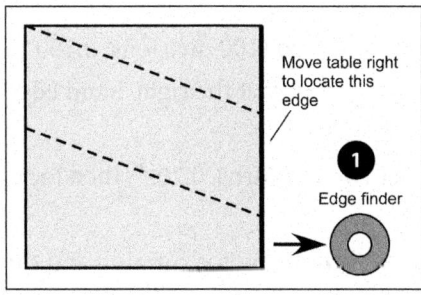
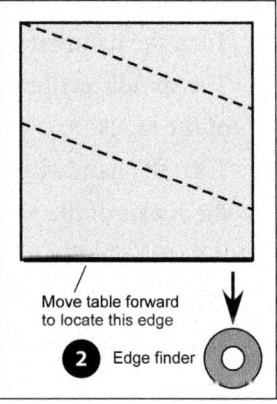

FIGURE 6-24 Edge finding.

FIGURE 6-25   Locating the right-hand edge.

2. Back the table to the left, then right again *very slowly* to verify the tripping point (by comparing the dial readings). Raise the edge finder clear; then *zero* the handwheel dial (by tightening the dial lock screw, if applicable).

3. Turn the handwheel another full turn (100 divisions, 0.100"). The spindle centerline is now precisely at the right-hand edge of the block.

4. Turn the handwheel another two full turns, 0.200"; then lock the X axis of the table.

5. Referring to Figure 6-24 (2). With the motor running and the edge finder lowered, use the same procedure on the Y axis, bringing the table *forward* to place the spindle centerline over the front edge of the block.

6. Bring the table forward 0.541" by turning the Y axis handwheel an additional five full turns plus 41 divisions. This places the spindle over location A, see Figure 6-23. Lock the Y axis of the table.

7. Exchange the edge finder for a center drill. "Spot-drill" a small cavity at location A.

8. Using a 1/8" drill (or other size, depending on your guide pins), drill through the block at A.

9. Unlock the X axis; then move the table right 0.850" by turning the handwheel eight full turns plus 50 divisions. Relock the X axis.

10. Without moving the Y handwheel, zero the Y dial.

11. Unlock the Y axis; then move the table forward 0.309" (= 0.850" − 0.541") by turning the handwheel three full turns plus nine divisions. Relock the Y axis.

12. Reinstall the center drill; then spot-drill at location B (Figure 6-26).

13. Exchange the center drill for the same size drill used in Step 8; then drill through the block at location B.

FIGURE 6-26 Spot-drilling hole B.

6-20 MACHINING THE ANGLED SURFACE

To machine the angled surface:

1. Insert the guide pins into holes A and B; then set the workpiece in the vise, as shown in Figure 6-27.

FIGURE 6-27 Block resting on guide pins.

2. Mill off the top surface using any size of end mill. Figure 6-28 shows a 3/8" end mill—the same one that will be used to machine the slot—but a larger one would need fewer passes. Start with a shallow cutter depth (say 0.03") to see how the material behaves; then adjust the cutter depth experimentally. The final pass should skim just the last few thousandths off the upper surface, but removing no material from face CD (Figure 6-23).

FIGURE 6-28   Machining top surface and slot

6-21 MACHINING THE SLOT

There are two concerns here: One is that the slot has to be centered (as in the three photos in Figure 6-23); the second concern is that the slot depth has to be 0.4".

To machine the slot:

1. First, be sure the workpiece is firmly held in the vise; then *remove the guide pins.*

2. Using the edge finder, position the spindle centerline over the front edge (indicated by the arrow in Figure 6-28). Zero the Y axis dial, then install the 3/8" end mill.

3. Using calipers, measure the width from the front surface to the back surface.

4. Bring the table forward by exactly half the measured width. If you started with the suggested 1-1/4" cube, this will be 0.625", given by six full turns of the Y hand wheel plus 25 dial divisions. Lock the Y axis. (To be sure the spindle centerline

is midway front to back, lower the quill to place the end mill on the upper surface; you can then use the caliper's depth rod to compare the distance from the cutter's OD to the front and back surfaces.)

5. To enable the cutter depth to be measured, lower the end mill gently to the surface, using the fine down feed; then zero the quill DRO, if available. If this is a knee mill, lock the quill, and slowly raise the knee to bring the surface up to the end mill. (Every machinist has a favorite way of doing this: Some like to place a very thin shim under the cutter, continually tugging on the shim to test for movability. When the shim is no longer free to move, the cutter is *approximately* even with the surface.)

6. Run the mill, and then traverse the X axis from side to side, with progressively increasing cutter depth (but first be sure the Y axis setting is correct by measuring the slot position after the first "just-skimming" pass—this, provided it's very shallow, won't be noticeable if the Y setting has to be adjusted). *Lock the quill before each cutting pass.*

7. When the slot depth is about 1/4", use calipers to get an accurate reading. If the measured depth is consistent with the quill DRO reading (or knee position), you can confidently continue machining until just short of 0.4". Check again; then make finishing cuts as necessary.

6-22 DRILLING THE THROUGH-HOLE

To drill the through-hole:

1. The centerline of the 1/4-20 tapped hole is 0.8" from point C in Figure 6-23. The snag here is that to use C as the datum point, we will have to unlock the Y axis and move the table forward so that the edge finder can touch the rearmost "ear"

(table moving to the left). Having zeroed the X dial at that point, you can move the table another 0.8" to the left, eight full turns.

2. We now need to reposition the spindle over the center of the slot exactly as we did in Section 6-21, Step 4. (Edge-find the front edge; then move the table forward by exactly half the measured front-to-back width of the workpiece. Lock the Y axis.)

3. Using a center drill, spot-drill at point E (Figure 6-23).

4. Drill a 13/64" through hole. Leave the drill in the chuck. (The by-the-book hole size for a 1/4-20 screw thread is #7, a difference of 2 mils; 13/64" was chosen because that's the nearest size you'll find in the typical home shop.)

6-23 COUNTERBORING THE UNDERSIDE

To counterbore the underside:

1. Reinstall the guide pins in A and B. Turn the workpiece upside down to allow point F (Figure 6-23) to be counter-bored. Tighten the vise.

2. Reposition the X axis to align the 13/64" drill with the hole just drilled. (*If the drill doesn't go in, was the hole properly centered front to back?* If necessary, adjust Y.) Lock the X and Y axes.

3. Install the 3/8" end mill in a holder or collet. With the spindle running at around 250 rpm, lower the quill (or raise the knee) *very slowly* until the mill is *just beginning* to cut a full circle. At that point, zero the quill DRO (or knee dial); then proceed to an additional depth of 0.270". If the socket head on your 1/4" screw is too tight in the 3/8" counterbore, ease the fit with a 25/64" drill.

6-24 THREADING THE THROUGH-HOLE

Reinstall the chuck; then thread the 13/64" through hole with a 1/4-20 tap. See Chapter 5, Section 5-44, for threading options; one way is to hold the tap in the chuck, hand-turn three-plus revolutions (to ensure proper alignment), and then finish with a tap wrench.

6-25 CUTTING OFF THE SURPLUS

To cut off the material at face G, reinstall the 3/8" (or larger) end mill. Set the workpiece in the vise, as in Figure 6-29, squaring it against the back vise jaw (or against the vise ways below the jaws).

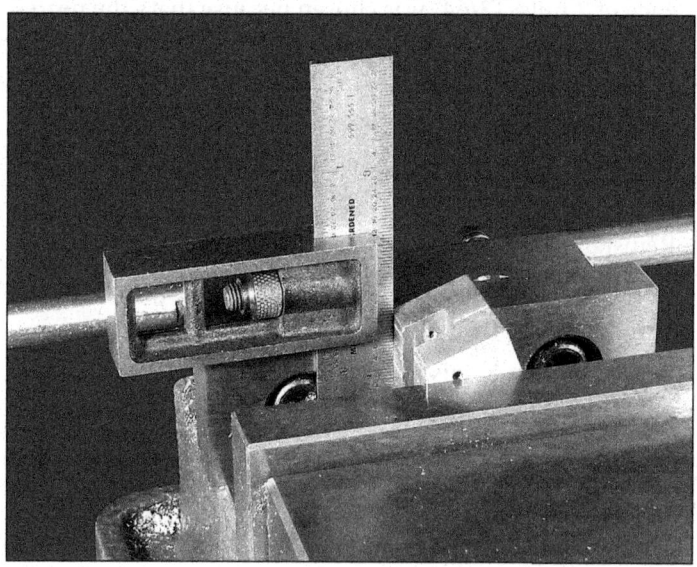

FIGURE 6-29 Squaring to cut face G.

CHAPTER 7

The Digital Readout

CONTENTS AT A GLANCE

7-1 THE BACKGROUND

The appearance of digital readouts (DROs) in the 1960s caused a revolution in machine tool thinking. For the first time ever, machinists could know—in real time—the *precise* position of a workpiece on the milling machine, or a cutting tool on the lathe.

Now, more than 60 years later, digital position displays of every imaginable type are within reach of the "occasional machinist" in both the product development lab and the home metal shop. Just a few minutes— even seconds—of hands-on experiments with a DRO-equipped mill will show you why the DRO is at the top of the machinist's must-have list. From the moment you switch it on, the DRO eliminates every element of "counting turns and divisions" and "one-way-only" motion to deal with backlash.

A mill DRO installation comprises a display unit, plus at least two scale assemblies permanently installed on the X and Y axes. There is an important difference between the two installations. The X axis scale *moves with the table*, as it traverses left to right; the reading head, which signals table position to the display, is *stationary*, fixed to the saddle (Chapter 1, Figure 1-9). On the other hand, the Y axis scale is stationary, fixed to the mill base (or the side of the knee); in this case, it is the *reading head that moves* forward and back with the table. In many cases, a third scale (Z) is installed on either the headstock or the knee. (This should not be confused with the quill DRO, a completely independent option on many bench mills.)

The X, Y, and Z scales were at one time always glass, precisely etched or otherwise marked to report movement with a resolution of one micron (about 0.00004"), some even less. Some DRO scales today are said to be "solid state," meaning they use a magnetized ferrite strip instead of glass. These scales are less fragile than glass and seem to be more forgiving of a less-than-perfect installation. Most milling machine scales have a *resolution of five microns*, approximately 0.0002". (Lathes usually have a one-micron scale on the cross-slide and a five-micron scale on the saddle.)

If you are installing your own scales, whether glass or ferrite, be prepared for a good amount of exacting work. This will include drilling and tapping cast iron, plus the fabrication of assorted brackets and scale covers (the hardware that comes with the scales can be helpful, sometimes, but there is always a need for modifications).

7-2 THE DRO DISPLAY

Figure 7-1 shows a generic three-axis display, with a calculator and the basic controls found on practically every DRO. Not shown are the preprogrammed macro routines that are usually called up by soft keys. Examples of such macros are bolt-circle drilling (also called "Pitch Circle Diameter"

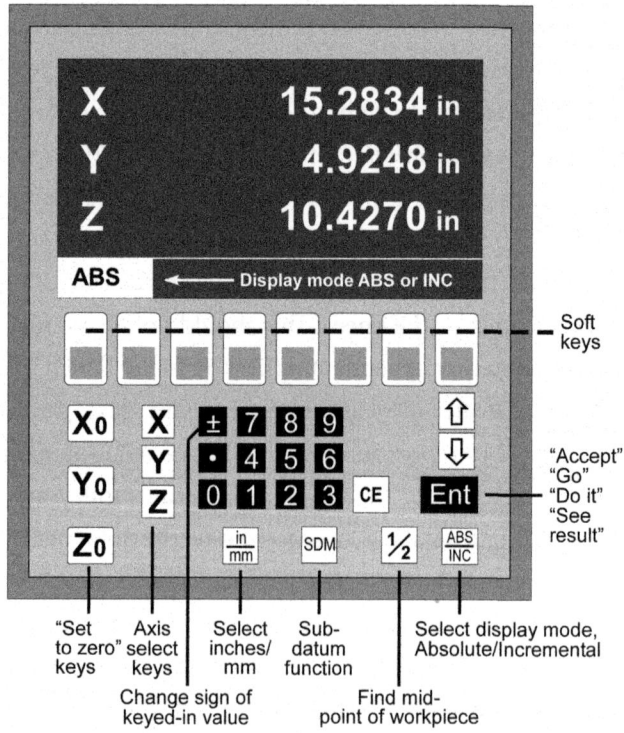

FIGURE 7-1 Typical DRO display.

drilling), holes in line, slot cutting, and arc cutting. These last two are "nibbling" programs that direct the user to move the mill table in very small increments to give a near-smooth edge (you might think of this as user-assisted CNC).

7-3 GRAPHICS OR NOT?

The defining difference between mill DROs is this: The more capable units come with a display that shows *graphically*, in real time, the mill's X and Y positions relative to the macro program, leaving no doubt about what step comes next (see Figure 7-12, shown later in the chapter). Practically all mill DROs have macro capabilities, but many have no graphics display. In other words, you may be flying blind, relying on numerical values and maybe a hand-drawn sketch of the program. *How important is that?* For many users, it isn't. Unless you foresee a heavy load of macro work, the extra cost may not be justified. For the odd occasion when you need to run, say, a bolt circle program, you can get by just fine with a non-graphics DRO.

7-4 KEYPAD

All DROs have a numeric keypad, which has three separate functions:

1. Keying in values that are called for in macro programs, e.g., X and Y values to define a starting point, radius of an arc or circle, angular separation.
2. Keying in "go-to values" to specify a desired displacement (offset) from the table's current position.
3. Math calculations, including add, subtract, multiply, divide, square, and square root. Also probably included are the trig functions sine, cosine, and tangent, plus inverse functions (e.g., arcsine). There is usually no memory. A $10 hand-held calculator has about the same functionality and in most cases is easier to use.

7-5 COORDINATES DEFINE LOCATIONS ON THE WORKPIECE

In everyday life, we use coordinates to specify a location on a map—for instance, *how far east is it from a point of reference, and how far north?* In milling machine terms, think of the map lying flat on the table, with south toward you. Westerly or easterly travel on the map corresponds to the longitudinal motion of the table, the *X axis*. Similarly, the north-south *Y axis* corresponds to the other direction of table movement, toward you, or back.

An X, Y coordinate pair completely defines a point in the plane of the mill table. If the project is simply to drill holes in a flat workpiece, all you need are X and Y pairs, as many pairs as there are holes. But sometimes, we need the "third dimension" to specify the depth of the holes or the vertical distance between two horizontal surfaces. This is the *Z axis* dimension, which has to do with the height of the headstock relative to the table.

A second Z factor is quill position—think "drill-press action." This is not reported on a three-axis DRO, because quill position is totally independent of headstock height (some bench mills have a separate quill DRO in the headstock).

7-6 POSITIVE- OR NEGATIVE-GOING VALUES WHEN THE TABLE MOVES?

In Figure 7-2, the X value becomes more positive as the table moves to the left; similarly, the Y value becomes more positive as the table moves forward, toward you. For everyday work, this is entirely a matter of choice, not an industry-wide convention (there isn't one).

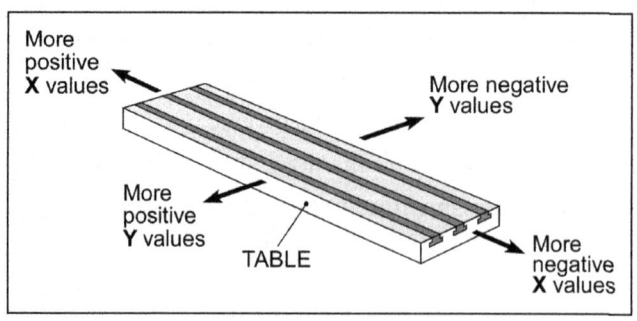

More positive **X** values

More negative **Y** values

More positive **Y** values

TABLE

More negative **X** values

FIGURE 7-2 A suggested sign convention.

The only time to be concerned about *display direction versus table motion* is when you are running a preprogrammed macro routine, in which case you hope there's guidance from the DRO manufacturer.

7-7 THE COORDINATE SIGN CONTRADICTION

Coordinate sign is a source of much confusion. As noted above, Figure 7-2 shows the displayed value becoming more positive when the table moves to the left. *Why is this, when we usually think of a leftward motion as "becoming less"? Shouldn't it be the other way around?*

The answer is *no*. Forget table motion as viewed by you in the ordinary way, facing the machine. Imagine instead that you are looking down the spindle centerline at the cutting tool as it travels across the workpiece. From this viewpoint, a left-moving table has the tool in effect moving to the right, which is the conventional positive direction in coordinate geometry (Figure 7-3).

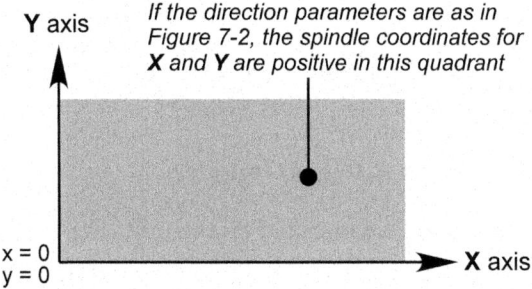

FIGURE 7-3   Looking down on the cutting tool.

7-8 RELATING COORDINATES TO REAL-LIFE MEASUREMENTS

In any discussion to do with "direction" and "coordinates," we always need to be specific about which is relative to what—see Section 7-7, above. For example, your sense of positive or negative can depend on your choice of workpiece datum—left side or right side, front or back (see Figure 7-4).

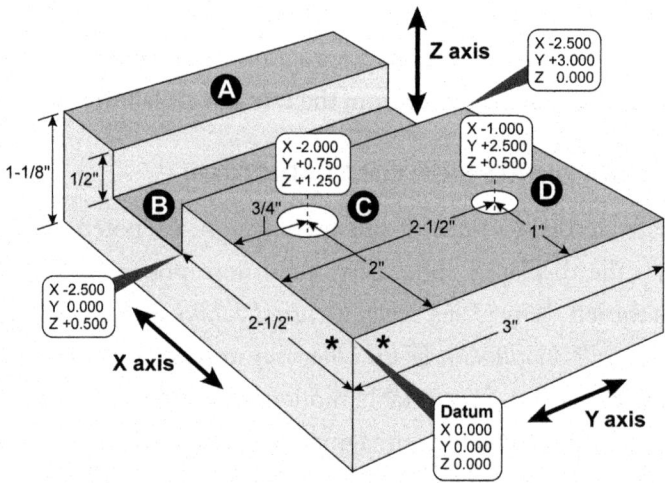

FIGURE 7-4 Real-life measurements and coordinate equivalents. By using an edge finder on the surfaces marked with an asterisk, the lower right corner of the workpiece has been defined in this example as the datum, the point where the X and Y values are both zero. The Z axis is assumed to be zeroed at the point where the cutting tool just touches the workpiece. The Z coordinates shown here indicate that the Z display goes positive—a user preference—as the cutting depth increases from surface A to surface B. Hole C is a through-hole, Hole D is not.

The two holes have negative X, but positive Y coordinates. Referring to Figure 7-2, this is because positioning the spindle over these locations calls for the table to be moved to the right (X) and brought forward (Y). If the DRO is set up with a different sign convention, the sign of one or both of the X and Y values will also change. This also applies if the datum is chosen to be at a different location, such as the lower left.

There are two other variables to consider:

1. How the scales were installed on the mill. The Y axis scale could be on either the left or right of the saddle. Ditto the Z axis scale, left or right of the column or knee. The X axis scale is mostly—but not in every case—at the back of the table.

2. No matter how the scales were installed, the way the DRO handles electrical output from the scales is up to you. *In every DRO, there is a setup routine for this*—look for a parameter labeled "axis direction" or "left/right," etc.

What it comes down to is this: For routine shop work, go with whichever direction setup you prefer. It really doesn't matter, unless you are using macros or sub-datums (Section 7-12). For these actions, you need to use the DRO manufacturer's sign convention, which may not be the one in Figure 7-2.

Another setup item to look for is actual versus displayed table motion; if the actual motion is 1", but the DRO reports 5", this may mean that one-micron scales are installed instead of the expected five-micron scales.

7-9 CHANGING MEASUREMENT UNITS AND ZEROING THE DISPLAYS

Most DRO displays have these basic functions (see Figure 7-5):

- **Choice of measurement units.** Press the in/mm key (1) at any time to switch from inches to metric, and vice versa.
- **Instant zeroing.** To set the current position of the X axis to zero, press the X_0 key (2). The Y and Z axes are zeroed in a similar way, keys Y_0 and Z_0.
- **Choice of Absolute or Incremental modes** (3), Section 7-10.

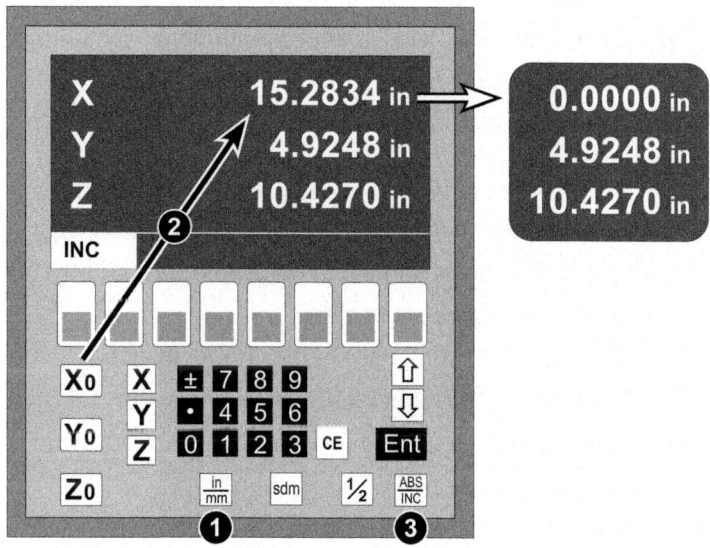

FIGURE 7-5 Basic DRO functions: (1) measurement units, (2) display zeroing, and (3) choice of mode.

7-10 CHOICE OF ABSOLUTE OR INCREMENTAL MODES

Note: The usual shorthand for these terms is "ABS" for "absolute mode" and "INC" for "incremental mode." Switching from one mode to the other is usually done with a single keystroke.

Superficially, the ABS and INC modes seem to be similar—both display X, Y, and Z coordinates in the same way, and the displays in either case can be zeroed in the ordinary way—see above.

The ABS coordinate frame is generally thought of as being fixed relative to the workpiece datum. The INC coordinate frame is set arbitrarily at any time in the machining process; in other words, INC has no fixed relationship to ABS.

Some users are content with whatever mode the DRO happens to be in when switched on. Many users set up ABS coordinates at the beginning of a work session, and thereafter work in the INC mode, knowing that the starting coordinates *can be recalled at any time* simply by switching back to ABS—highly recommended.

The following example illustrates one way of using a combination of the two modes:

1. Select a workpiece datum. In Figure 7-6, the bottom right-hand corner has been "found" in the ABS mode, and the XY coordinates have been set to zero.
2. In ABS mode, drill hole A at X – 0.350, Y + 0.300.
3. For hole B, move the table 1.15 right to display X – 1.5 (the sum of –0.35 and –1.15) and forward to Y + 0.6 (the sum of 0.3 + 0.3). Drill hole B.
4. The group of smaller holes C, D, and E is located by reference to hole B. One possibility is to stay in the ABS mode and

move to C, D, and E using mental arithmetic—but it's easier to switch to INC mode, zeroing X and Y at the B location.

5. In INC mode, hole C is +0.5 away from B in X and +0.25 away in Y. Drill hole C, and then zero X and Y at C before moving on to D and E.

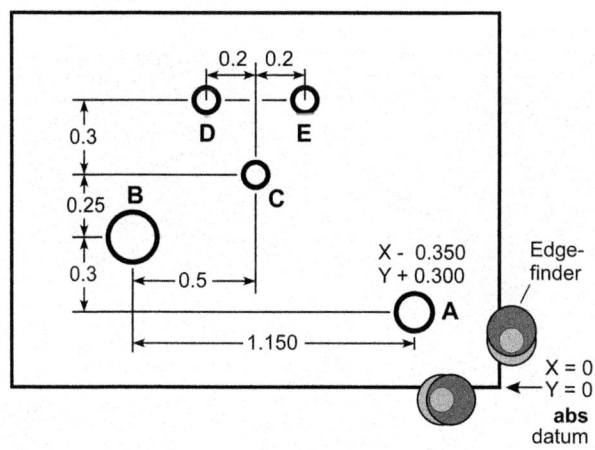

FIGURE 7-6 Using ABS and INC modes.

None of the above affects the ABS datum, so the starting point can be recaptured at any time.

7-11 FINDING MIDPOINTS AND CENTERS

Locating the exact center of a workpiece is one of the most used applications of the DRO. The examples in this section show a conventional edge finder with a spring-loaded tip; when the tip just touches the workpiece edge, it kicks out suddenly—edge "found." It helps to use a high spindle speed for edge finding.

The following is a typical procedure. It may not apply exactly to your DRO, but the principle will be the same.

To find the center of a rectangular workpiece:

1. With the DRO in INC mode, and with the edge finder positioned as shown in Figure 7-7 (A), bring the table slowly forward to the point where the edge finder tip kicks out at the leading edge of the workpiece.

FIGURE 7-7   Finding the Y axis center on a rectangular workpiece.

2. Zero the Y axis.
3. Raise the edge finder clear of the workpiece.
4. Without disturbing the table's X position, bring the table forward, clear of the workpiece.
5. Lower the edge finder as shown in Figure 7-7 (B).
6. Move the table away from you to "find" the back edge of the workpiece; then press the 1/2 key, followed by the Y axis select key.

The Y axis value is now *one-half* of the workpiece's front-to-back dimension (plus two times the edge finder's tip radius). Now move the table backward until the Y display reads 0.0000. At this point, the spindle is exactly over the midpoint between the front and back edges.

Repeat the above steps at the left and right edges of the workpiece to find the X axis midpoint.

To find the center of circular features: Locating the midpoint as described above is not only for rectangular objects. It can also be used to find the center of circular objects. Examples include (1) locating the center of a circular bar, maybe to center-drill it for lathe work (Figure 7-8), and (2) locating the center of a hole (Figure 7-9).

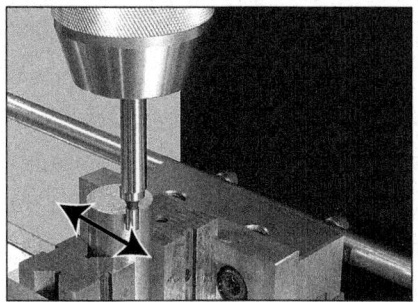

FIGURE 7-8 Finding the X axis center on a circular bar. Be sure the Y axis location of the table is exactly the same when finding the left and right edges of the bar. The workpiece must be truly circular for a reliable result.

FIGURE 7-9 Finding the Y axis center of a circular hole. Be sure the X axis location of the table is exactly the same when finding the front and back edges of the hole.

Locate the edge finder as near to the estimated centerline as possible before touching the edge. It does not have to be exactly on center, but the "other axis" *must not move* when you traverse from one side of the feature to the other—clamp it if necessary.

7-12 SUB-DATUMS

Most DROs provide ways to set up a number of secondary ABS coordinate frames known as "sub-datums"—in some cases, a hundred or more. They behave in exactly the same way as the "native" ABS and *are referenced to the ABS datum.*

> This means that if ABS is re-zeroed at a new location, all sub-datums are moved in lockstep.

Sub-datums, or SDMs, are helpful in machining operations calling for more than one point of reference. If the sub-datum function did not exist, "virtual datums" could be established by zeroing in the INC mode at any time in the machining process. The shortcoming of that approach, compared with using true sub-datums, is that "DIY" virtual datums are not are not related to each other in a specific way, nor are they related to the ABS datum, other than by the offset numbers (which maybe you will have remembered to make a handwritten record of, or maybe not?).

How sub-datums are established depends on the DRO design, but most use some variant of what is sometimes referred to as the "Learn method." The basic idea is to select the next available (unused) number in the SDM register; then move the table to bring the spindle over the location you would like to define as the new SDM. It is usually that simple.

In *some* DROs, sub-datums can be used just like ABS for all preprogrammed routines. *Caution:* This doesn't apply to all DROs.

Figure 7-10 shows a workpiece with the ABS datum at bottom left and three subdatums. The following is a generic procedure for setting up the values. The sign convention is as shown in Figure 7-2. Your DRO may be quite different, but the end result should be similar.

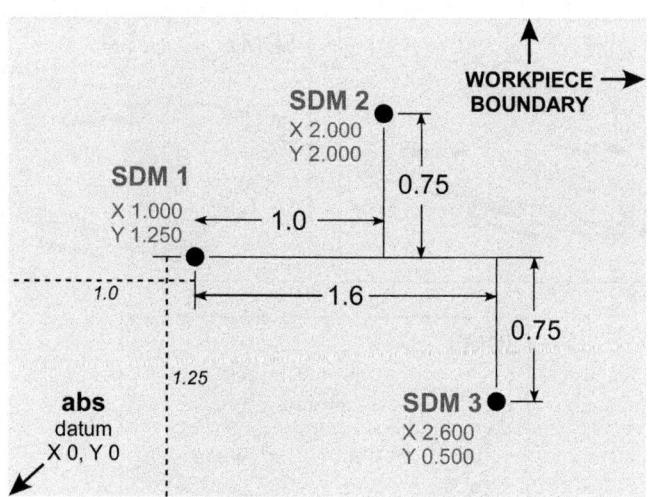

FIGURE 7-10 SDM example.

1. Select ABS mode.
2. Establish the datum by edge finding or other means; then zero the X and Y values, Xo and Yo.
3. Move the table to the SDM 1 location—left to X 1.0000, forward to Y 1.2500.
4. Press the SDM key. Typically, you will be asked for an SDM number. Enter 1 on the keypad; then press ENT to set SDM 1.
5. Move the table to the SDM 2 location—left to X 2.0000, forward to Y 2.0000.
6. Press the SDM key again. Enter 2 on the keypad; then press ENT to set SDM2.
7. Move the table to the SDM3 location—left to X 2.6000, back to Y 0.5000.
8. Press the SDM key again. Enter 3 on the keypad; then press ENT to set SDM3.

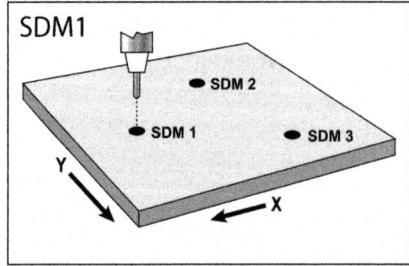
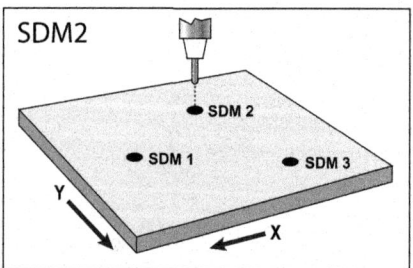
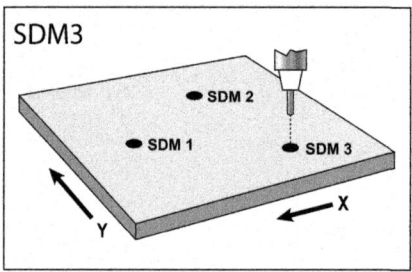

FIGURE 7-11 Setting up SDMs.

Using the sub-datums in a machining project is typically a matter of pressing the SDM key, then entering the desired SDM number on the numeric keypad. The values then displayed by the DRO are *with reference to the just-selected SDM datum*, which is in turn referenced to the *unchanged ABS datum*.

7-13 MACHINING PROGRAMS (MACROS)

Most DROs offer a number of programs that can automate frequently used operations in the machine shop. The most popular of these are three bolt-hole drilling programs (also known as PCD programs): (1) full circle divided into a specified number of equal sectors; (2) arc of a circle, divided equally, start and end angles specified; and (3) custom arc of a circle, sector angles individually specified. Not all of these options are available in every DRO.

Programs for drilling holes in lines and grids are usually included, plus various *contouring macros*, used to "nibble" a series of overlapping holes along a line (slot cutting) or around the perimeter of a rectangle or arc. Smoothness of the nibbled outline is usually controllable by a "maximum-cut" parameter, which limits the size of step between one hole center and the next—the smaller the maximum step, the smoother the outline.

Some DROs have a *graphics display* that shows where you are in the program (Figure 7-12). Not essential, but it does simplify the process.

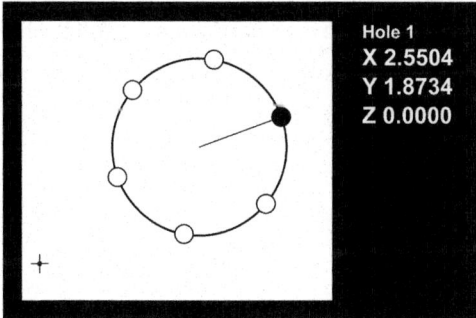

FIGURE 7-12 Example graphics display, six-hole bolt circle. The ABS zero (+) is at bottom left in this example. "Distance-to-go" values are at top right—when these values are zeroed, the spindle is at hole 1. This macro is defined by four parameters: number of holes, center location, pitch circle diameter, and start radius angle (20° here).

FIGURE 7-13 Do a dry run to avoid mistakes. It takes only a few minutes and wastes only scrap material, not the whole project. This shows rough locations for an 8-hole bolt circle.

BEFORE CUTTING METAL

This is a caution for all macro routines:

Check that the hole configuration calculated by the DRO is as expected. Always do a dry run by marking the rough coordinates of the holes on the workpiece with a fiber-tip pen; then run the macro for comparison—maybe even doing a trial run on scrap material (Figure 7-13). If the hole layout is flipped in either the X or Y axis, or both, it may be that the axis direction parameters need attention.

7-14 BOLT-HOLE DRILLING

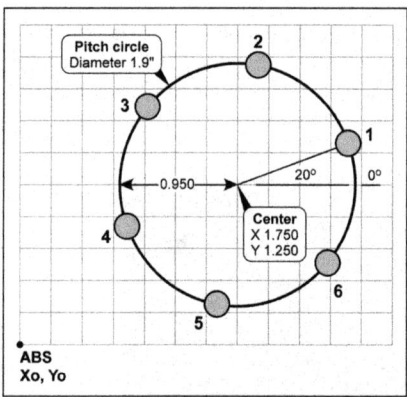

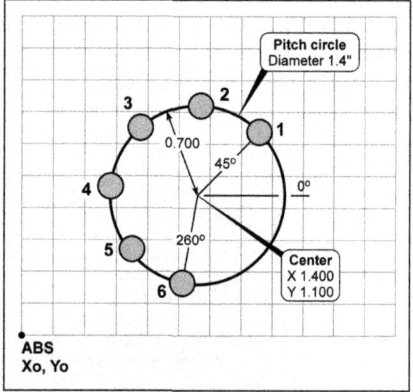

FIGURE 7-14 Complete circle. Six equally spaced holes, start angle 20°.

FIGURE 7-15 Arc of a circle. Six equally spaced holes, start angle 20°, end angle 260°.

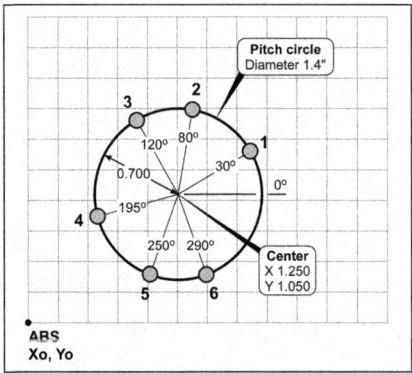

FIGURE 7-16 Custom pattern. Six holes, start angle 30°, end angle 290°, separations individually specified.

7-15 HOLES-IN-LINE DRILLING

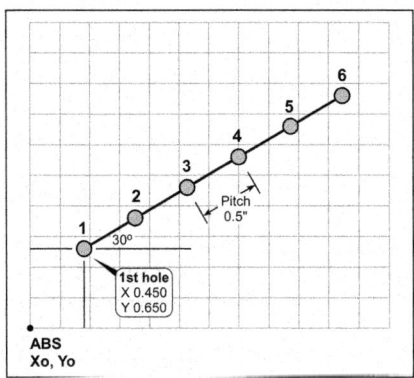

FIGURE 7-17 Custom pattern. Six holes, start angle 30°, pitch 0.5".

7-16 SLOT CUTTING

Slot smoothness is controlled by a selectable limit on step size between holes (Figure 7-18). *For nibbling, use a center-cutting end mill instead of a drill.*

7-17 ARC CONTOURING

Smoothness is controlled by a selectable limit on step size between holes (Figure 7-19). Most DROs offer the choice of cutting on the circumference of the circle or touching the circumference inside or outside. *For nibbling, use a center-cutting end mill instead of a drill.*

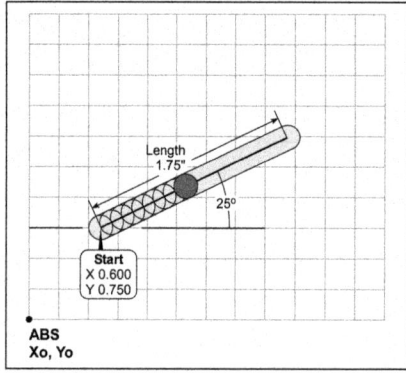

FIGURE 7-18 Slot specified by length and angle.

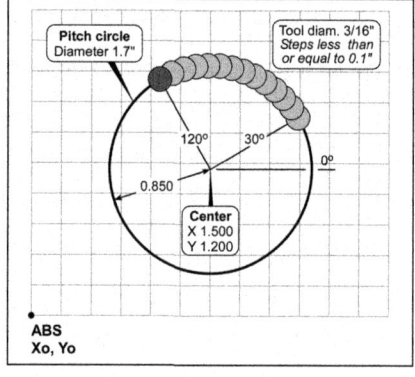

FIGURE 7-19 Arc contouring. Specified by start and end angles, cutter path on-the-line or touching the line, internally or externally.

The Rotary Table

CONTENTS AT A GLANCE

This chapter assumes that a DRO is installed—not essential, but less time-consuming (and less error prone) than dead reckoning.

8-1 EXAMPLES OF ROTARY TABLE WORK

(A) GENEVA MECHANISM Gears 1 and 2, 60T and 30T, were machined using the same setup as in Example F. There was no need for a dividing head for 30T—rotate three full turns between each machining pass. Not so for 60T (1-1/2 turns), so a dividing plate was required. (This is one case where a spin indexer might be more efficient—see the box at the end of the chapter.)

(B) MACHINING A GENEVA WHEEL This is similar to the Geneva wheel in Example A. The setup here gives some idea of what to expect in rotary table work—there is usually a fair amount of custom fixturing, such as the dowel pin (which the arrow is pointing to in the figure), plus the circular series of holes (which were drilled using a standard DRO program, not the rotary table). The table surface is protected by 1/4" PVC sheet.

(C) MILLING HEXAGON FLATS ON 5/8"-DIAMETER STOCK For this application, the chuck is rotated in 60° increments, so there is no need for a dividing plate—simply turn the crank 15 full turns each time. The regular handwheel could be used here instead of the crank handle.

(D) BLENDING CURVES, MILLING SLOTS

(1) Steel, 2" across the flats, a 1"-radius semicircle blended to machined flat surfaces either side. (2) Aluminum, 5" diameter, three 1/2" x 1/4" diametrical slots at 60° spacing. (3) and (4) Three-lobe support plates for ball bearings—brass, each lobe threaded for a bearing stud. Smooth blending of the 0.2"-radius lobes with the much larger main arcs (inset).

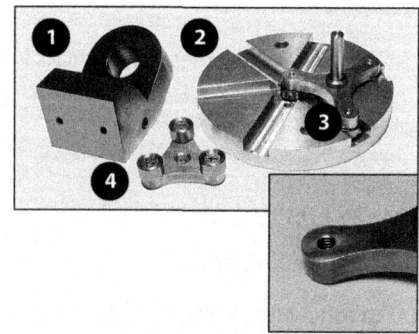

(E) CUTTING A 1-1/2"-LONG, 1/16"-WIDE SLOT

This is one of six butterfly control shafts for a fuel injector system—an example of precise work using the tailstock. In addition to the slot, one end of each shaft was squared, then threaded on the lathe for the drive linkage.

(F) CUTTING A 34-TOOTH GEAR

This application called for a dividing plate with a 17-hole circle. The interval between teeth was set by two full turns of the crank, plus 11 additional spaces. To eliminate the need to count holes on the dividing plate, "sector arms" (indicated by the arrows) are set to span the 11 spaces. After each machining pass, the sector arm assembly is swung around manually to set up for the next two-plus 11 cranking operation.

(G) MILLING A LARGE COUNTERBORE

This is a 1.4"-diameter, 0.15"-deep counterbore in aluminum. It was cut in three consecutive passes, with progressively smaller diameters. Exactly the same setup could be used to cut a through hole. Note the MDF beneath the workpiece.

(H) DOG CLUTCH FOR MACHINE TOOL

Steel, 1.75" diameter. The spec called for a 320° ± 0.5° circular slot, 1/4" wide.

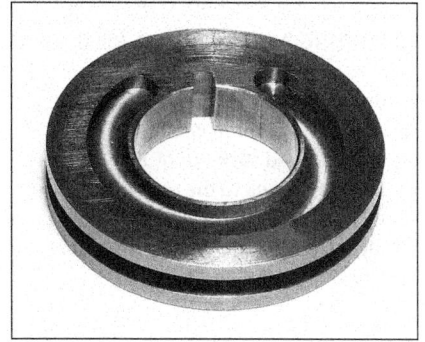

(1) CUTTING AN INTERNAL O-RING GROOVE   This is a 1/16" x 1/16" groove cut in a section of precision seam-welded stainless steel tube, 1-1/4" ID. (Hardness of the seam made this difficult to machine on a lathe.) The rectangular HSS cutter bit here was ground from an old #40 drill. It was held in a 3/8" cutter shank (drill rod) by two #4-40 set screws, with a measured stick-out of 1/16". This allowed groove depth to be gauged by proximity of the cutter shank to the tube wall. After each full turn of the rotary table, the X position of the main table was adjusted to close the gap progressively to zero (cutter shank rubbing gently on the ID). The spindle speed was 1,500 rpm, and the rotary table was slowly turned counterclockwise.

8-2 THE BACKGROUND

About 80% of milling machine work is metal removal in straight lines, usually at right angles. Another 15% or so might be drilling, boring, and thread cutting. Whatever the percentages, once in a while the job cannot be completed without some means of cutting circular arcs. There are two choices for dealing with this. Today, one suggestion you might hear is "Convert your mill to CNC" (computerized numerical control), or pass the job along to someone who already has that capability. The other choice is to add a rotary table to your mill—much less of a stretch than CNC conversion, but a challenge in its own way. This is because the rotary table isn't a simple add-on accessory. It adds (literally) a new dimension, *angular measure.* This allows you to machine, with high accuracy, practically any kind of arc-shaped or full-circle feature. The downside of the rotary table is that it calls for unfamiliar math, as well as a collection of hold-down and positioning accessories unlike those used on the basic mill. However, most of such accessories can be made in the shop.

This chapter describes a *generic* 6" rotary table (Figure 8-1)—"generic," because it has been copied by so many manufacturers in the Pacific Rim, and elsewhere, that no one is telling where it originated. (For instance, I have two 6" tables, one made in Taiwan, the other in India; they are practically identical—the dividing plates, probably from China, fit both tables.) A defining feature of this style of rotary table is its *drive ratio*, in this case *90:1*, which means that one turn of the handwheel turns the table 4°.

FIGURE 8-1   A 6"- diameter (150-mm) rotary table. This is the standard handwheel configuration. If the table is used with dividing plates, the handwheel is exchanged for a crank arm, shown in the inset at lower left.

The rotary table can be installed on the mill table either horizontally or vertically. Two hold-down slots for up to 1/2" bolts are provided for each orientation. (If the handwheel touches a front-mounted DRO or other machine component in the horizontal setup, an intermediate mounting plate may be required.) This can also apply to the vertical setup if the T-slots on your mill table don't match the holes on the rotary table (Figure 8-2).

FIGURE 8-2 Vertical and horizontal installations of the rotary table, with optional 3-jaw chuck and tailstock. Use a precision square, inset photo, for accurate alignment with the mill table. For vertical mounting, the T-slots on the mill table have to match the holes in the rotary table base (see the dimensions, inset. If they don't match, use an intermediate aluminum plate, 3/8" or 1/2" thick. Tailstock and chuck are usually sold separately).

Product packages and specs vary somewhat, but tables of this size usually come with a Morse taper #2 (MT2) center hole and either three or four radial T-slots (if you have the choice, opt for four). The table is rotated by a handwheel and a 90:1 worm drive, so one revolution of the crank handle turns the table 360/90 = 4°. The handwheel is usually graduated in 2 minutes of arc intervals (2'), plus there is a vernier scale for enhanced resolution.

Three important options are sometimes offered (Figure 8-3):

1. A 3-jaw chuck and backing plate (the backing plate should have an MT2 stub to locate directly in the rotary table center).
2. A stand-alone tailstock.
3. A set of dividing plates to step the table around in 150 precise increments, without the need for calculation or vernier interpolation. *These are not available for all 6" tables—you need to ask.*

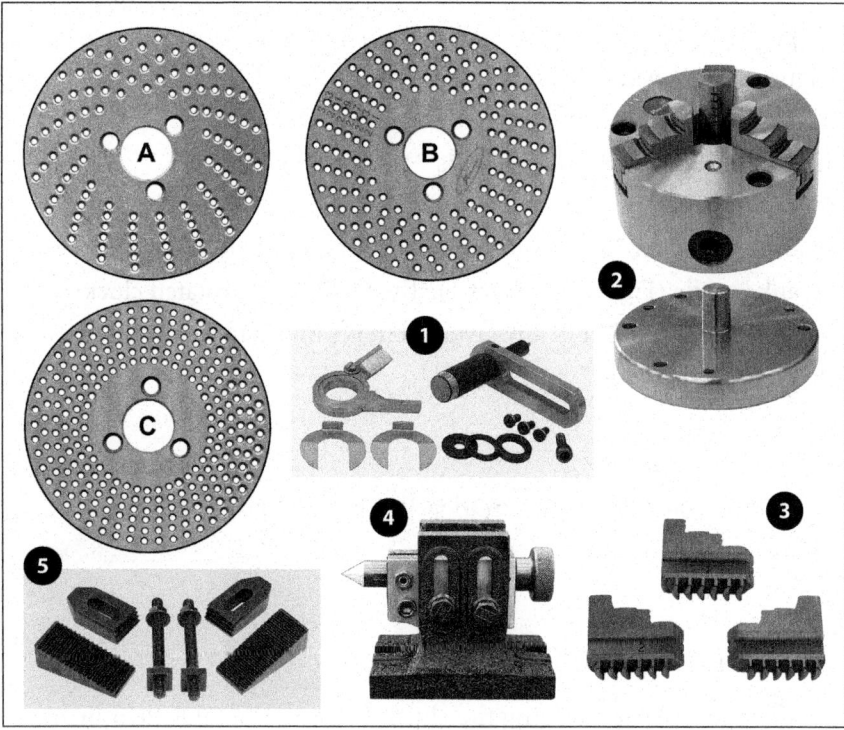

FIGURE 8-3 Typical accessories. A, B, and C: Dividing plates, 4" diameter, total of 18 hole circles. (1) Crank handle and sector arms for dividing plates; (2) 3-jaw chuck and backplate with Morse taper #2 stud, shown here inverted; (3) Second set of jaws for large-diameter workpieces; (4) Tailstock; (5) Hold-down kit.

The first part of this chapter describes conventional operations using the handwheel and vernier scales to set the table's angular position.

Procedures for setting the table using dividing plates are covered in the second part, starting at Section 8-14.

8-3 TABLE DRIVE MECHANISM

With the clutch lever OFF (Figure 8-4), the table can be freely rotated to any angular setting, then clamped in position by two table locks (Figure 8-1). When the clutch lever is ON, the table is usually turned by the handwheel, the dial of which is

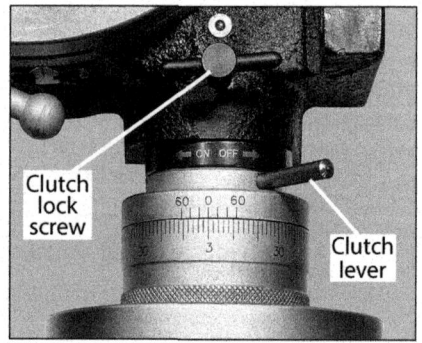

FIGURE 8-4   Rotary table controls. If the lever is obstructed by the mill table, the lever hub can usually be loosened and rotated to clear the table.

graduated in two minute-of-arc intervals (2'). When the table is used with dividing plates, for "mechanical indexing," the handwheel is exchanged for a crank handle (Figure 8-3, inset 1). The table can be rotated clockwise or counter-clockwise (but in any given job it is important to turn in the same direction throughout).

Be sure to tighten the lock screw when the clutch lever is set to ON.

8-4 TESTING AND ADJUSTING BACKLASH

Test for backlash by swinging the clutch lever first to OFF, then firmly to ON. Tighten the clutch lock screw (T-screw) above the eccentric sleeve (Figure 8-5). With the table held firmly with both hands, try moving it back and forth (be sure the table locks are off—see Figure 8-1). If movement is detected, swing the lever to OFF; then unscrew the backlash adjuster a few degrees (Figure 8-6). Reengage the worm—clutch lever to ON and locked—and then check again for movement. Repeat as necessary to minimize backlash, but not to the point where it is difficult to turn the handwheel smoothly.

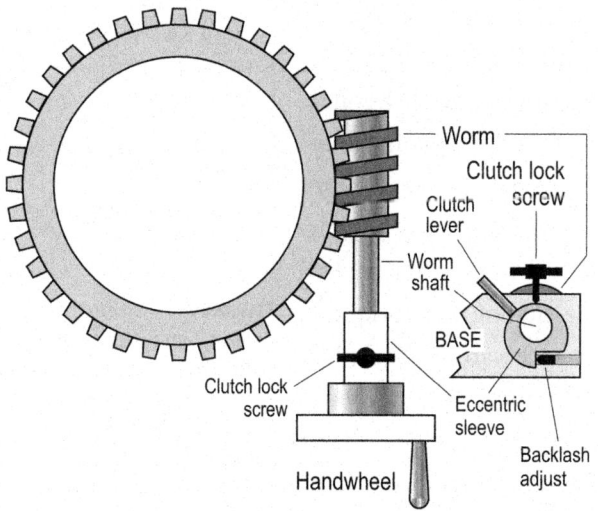

FIGURE 8-5 Rotary table schematic. When the clutch lever is moved to the left, counterclockwise, the eccentric sleeve rotates to bring the worm into contact with the worm wheel beneath the table. The clutch lock screw holds the worm engaged. To adjust the backlash, see Figure 8-6.

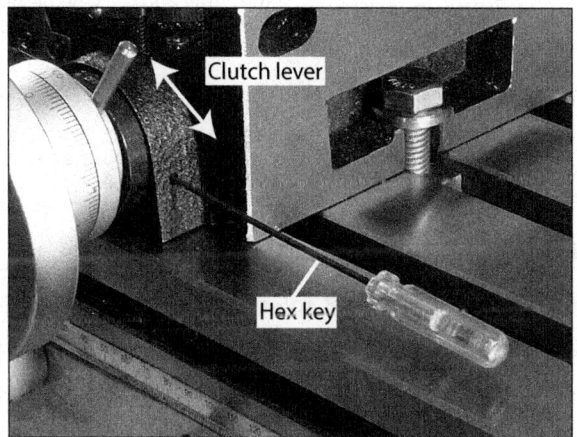

FIGURE 8-6 Backlash adjustment. To tighten the mesh, first swing the clutch lever clockwise to OFF, slightly unscrew the set screw, and then return the clutch lever to ON.

8-5 WHICH DIRECTION TO MOVE THE WORKPIECE?

In *conventional milling*, the cutter pushes the table against the positive thrust of the lead screw, as opposed to pulling it into the "backlash zone." In everyday X and Y axis machining, we can usually choose *conventional* versus *climb milling* simply by positioning the table to bring the cutter to one side of the workpiece or the other—forward or back, left side or right side. With a rotary table, however, we often don't have the same freedom of choice. Many rotary table projects call for a combination of conventional and climb milling on a single feature—such as the circular slot in Figure 8-7. Setting up for *minimum backlash* is therefore an important prerequisite for any project that may call for

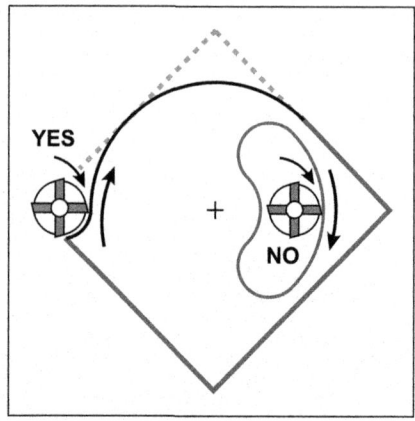

FIGURE 8-7   Normal milling versus climb milling. Two typical examples of rotary table machining are (1) corner rounding and (2) slot cutting. In conventional milling ("YES"), the cutter edges at the point of contact are traveling in the opposite direction to the workpiece. Backlash isn't a factor, because the cutter is pushing against the pressure of the worm drive. Not so when cutting the outer arc of the slot ("NO"), an example of climb milling. If backlash is present, the cutter may unexpectedly nudge the workpiece around, possibly damaging both the cutter and the workpiece. See Section 6-5 for other slot-cutting issues.

both clockwise and counterclockwise table rotation. Sometimes there is simply no option—so check for backlash before you start.

8-6 SETTING UP THE TABLE—ROUGH ZEROING

The usual starting point for most rotary table operations is adjusting the XY settings of the mill table to bring the table's center of rotation into alignment with the mill spindle. At this point, the DRO, if installed, is set to X = 0, Y = 0.

The process can be made easier and faster using a centering aid
such as shown in Figure 8-8 (it
also appears as adapter D in Figure
8-9). This was shop-made from a
stock item known (variably) as a
machinable blank, a hardened MT2
taper with a large-diameter parallel
(unhardened) shank. In the example shown in the figure, the parallel
portion was turned down to exactly
3/4" diameter by installing the MT2
blank in a taper-reducing socket in
the lathe spindle.

FIGURE 8-8   Positioning the table.

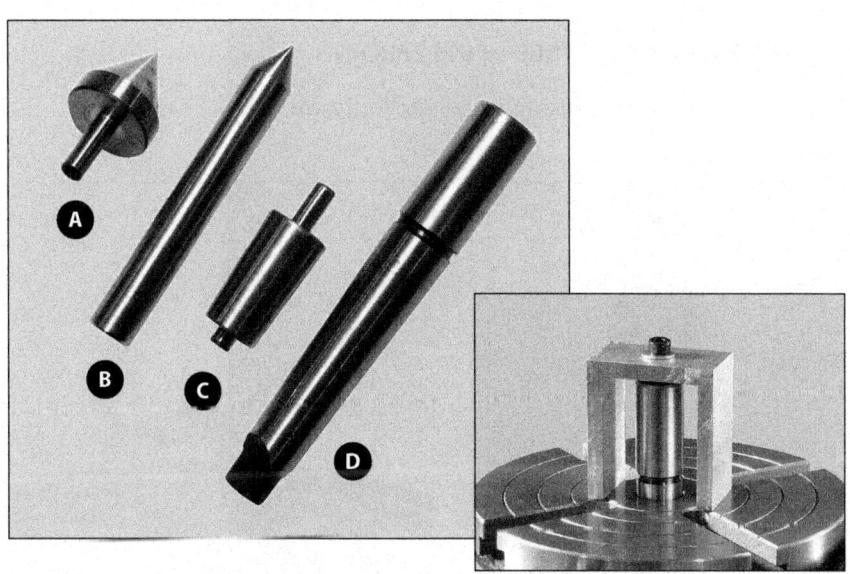

FIGURE 8-9   Centering aids. A and B are 60° points with parallel stems, 1/4" and 1/2",
respectively. C is a shop-made MT2 taper body with a 1/4" dowel pin. D is a shop-modified
MT2 taper with a 3/4" parallel extension. (D is internally threaded for a jacking screw,
shown in the inset photo. This ensures that it can be removed without access to the table's
underside.)

To use a centering aid like adapter D, set the rotary table on the mill table with attachment bolts in place. The bolts should be snug—but not fully tightened—allowing the rotary table to slide freely. Install the 3/4" parallel shank of adapter D in a collet (Figure 8-8); then adjust the mill table to position the adapter over the rotary table's tapered bore. Check by repeatedly lowering the quill, making increasingly fine XY adjustments to the point where the taper can be plunged full depth. The rotary table is now close to perfect alignment, so tighten the adjustment bolts *a little more*; then zero the DRO.

Do all of the above without clamping the milling machine quill at any time— this can cause a misalignment of several mils.

8-7 SETTING UP THE TABLE—FINE ZEROING

Make final adjustments using a dial test indicator (Figure 8-10).

FIGURE 8-10 Aligning the rotary table to the mill spindle. Any type of dial test indicator such as these will do the job. To minimize runout error, use a collet instead of a chuck. Rotate the mill spindle *by hand*, checking the reading as you go. A mirror helps when the indicator is facing to the back.

Set the tip of the dial test indicator in the tapered bore; then check for runout by rotating the spindle. If necessary, use a mirror for 360° visibility.

Whatever else you do, *don't fall for the classic trap* of cranking the rotary table around with the *mill spindle stationary*. This tells you how concentric the bore is, not how the table is positioned.

Make final adjustments to the X and Y settings as necessary; then zero the DRO again. If you have used the plug-in adapter D, as described, these final adjustments should be only a few mils.

If there's a frequent need to zero the rotary table, consider using a coaxial dial indicator; this will eliminate the need for a back-view mirror, and can save a lot of time (see Section 5-19).

8-8 READING ANGULAR POSITION

The outer rim of the table is graduated in degrees from 0° to 360°. This is the main angle indicator (Figure 8-11).

The handwheel dial has 120 divisions, each representing two minutes of arc (2'), shown in Figure 8-12. One revolution of the crank rotates the table 4°. The 60-0-60 vernier scale makes it possible to resolve to the nearest 20 seconds of arc (20").

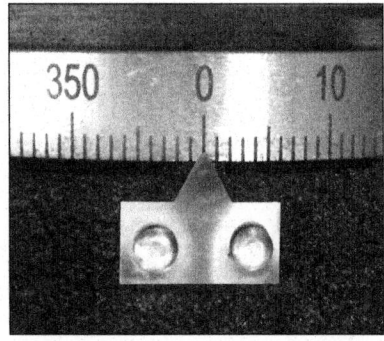

FIGURE 8-11   Main angle indicator.

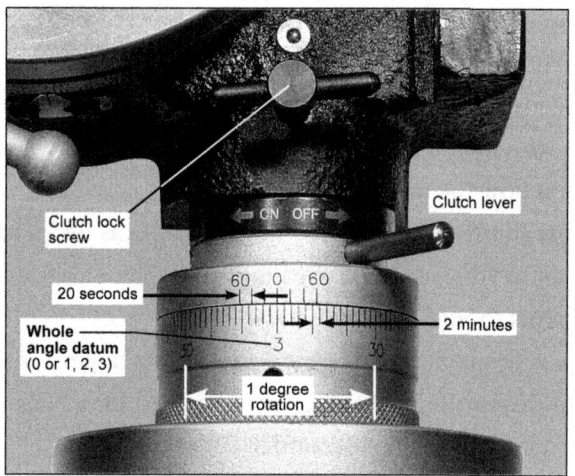

FIGURE 8-12 Dial graduations.

8-9 READING THE VERNIER

Referring to Figure 8-13:

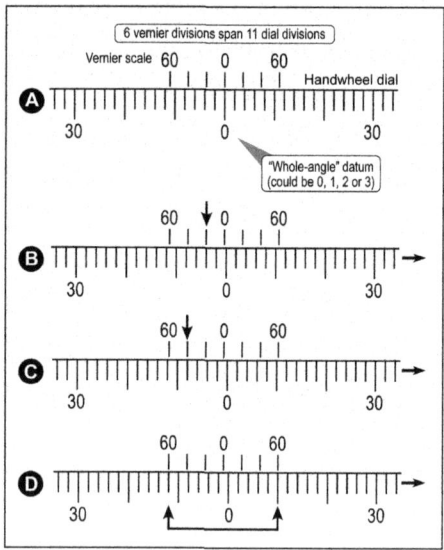

FIGURE 8-13 Reading the vernier.

A. The whole-angle datum 0 on the handwheel dial is aligned precisely with the vernier 0. No other lines coincide.

B. The crank has been turned 20 seconds clockwise (written 20"). You can see that the movement was exactly 20", a tiny amount, because only at that point (arrow) do the vernier and handwheel dial graduations coincide.

C. The handwheel has been turned a further 20 *seconds* clockwise, shifting the coincidence one notch along the vernier.

D. The handwheel has turned yet another 20 seconds, one whole
    minute beyond A, placing the vernier 0 midway between
    graduations on the handwheel dial. Note the coincidence at
    both 60 lines on the vernier.

8-10 USING THE ROTARY TABLE

All rotary table applications are variations of the following:

1. Cutting a circular path, an arc, of an exact number of degrees,
    using an end mill as in Example H, Section 8-1. An exten-
    sion of this is cutting a full circle to make a counterbore of a
    specified depth, Example G Section 8-1, or cutting through
    to make a large hole. (Take care when setting the radial offset
    precisely: Diametrical error is twice the radial error.)

2. Drilling a hole at a given radius from the table's center of
    rotation, at a specific angle from a workpiece datum. An
    extension of this is drilling a circular pattern of evenly spaced
    holes, as shown below in Figure 8-14. Section 8-12 describes
    a procedure for this, showing how to step the table around by
    counting a given number of turns of the handwheel, followed
    by fine-tuning using the vernier scale.

    With some rotary tables, you have the option of using the
    dividing plates instead, without the need for vernier fine-tuning;
    see Section 8-14.

8-11 A CIRCULAR-HOLE PATTERN DEMO

First, zero and lock the X and Y axes on the DRO (ABS mode) with the
spindle exactly in line with the table's center of rotation (Sections 8-6 and
8-7).

The circular-hole example described here has seven holes (Figure
8-14). Hole 1 is on the diameter at right angles to the side designated as the
reference edge. This is an important component of any rotary table setup.

Whatever drilling or cutting you are planning to do is almost always oriented with respect to a feature of the workpiece you can measure against—an edge, an existing row of holes, etc.

Here is the setup procedure for the seven-hole circle:

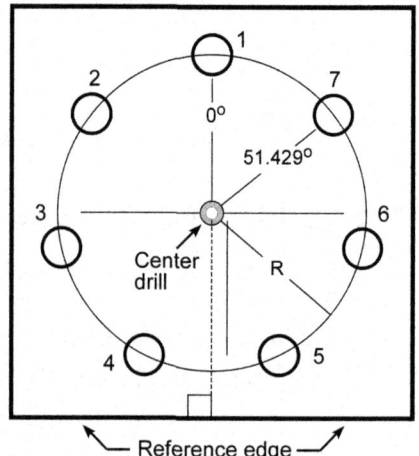

FIGURE 8-14   Seven-hole circle.

1. Start with a workpiece of scrap material about 3" square. Center-drill the workpiece. Also, as a sanity check to avoid gross mistakes, sketch out the pattern using a fiber-tip pen or dabs of marking fluid. You might be wondering, *On a real-life workpiece, what if center-drilling is not possible? How would you position the hole circle?* There is no good answer to that question, but in practice a tiny reference hole is rarely a problem, even if it's visible in the end product.

2. Position the center-drilled hole under the spindle's centerline (*which should still be at X = 0, Y = 0*). Figure 8-15 shows an easy way to do this, using the centering pointer B (Figure 8-9, shown earlier). Clamp the workpiece to the rotary table, leaving clear (for drilling) as many of the seven positions as possible.

   To drill all seven holes, you will probably need at some point to relocate the clamps *without shifting the workpiece*.

3. In the final part of the setup, align the reference edge with the mill's X axis, thus making sure hole 1 will be on the "right-angled diameter." The objective here is to achieve alignment

FIGURE 8-15 Drilling setup. Numbers 1 and 4 refer to Figure 8-14.

FIGURE 8-16 Aligning the reference edge.

when the table is at an *angle you can remember*, and therefore return to easily if you lose your way in the drilling sequence. Be prepared for some fine-tuning. Attach a dial indicator to the mill headstock using a magnetic base, or hold it in a drill chuck or collet (Figure 8-16).

4. With the rotary table clamps loose, and the clutch lever turned to OFF, align the reference edge to the mill table as well as you can by eye. Set the clutch lever to ON; then tighten the clutch lock screw to give control of the table to the handwheel.

5. Release the *Y axis lock,* only; then bring the workpiece forward to bring the indicator tip over the reference edge. Lower the headstock or spindle to make contact; then bring the workpiece forward a little more to preload the indicator. Traverse the X axis left to right. Work toward perfect alignment—stationary needle   by rotating the table in a series of small increments. When the indicator reading is the same at both ends of the workpiece, hole 1 will be correctly located relative to the reference edge.

> ***Important!*** Decide in advance whether the table will be going clockwise or counterclockwise when you are drilling the hole pattern—and be sure to *use only that direction* when aligning the workpiece (if you overshoot, back the handwheel a few degrees; then try again). *In the following notes, the rotation is clockwise.*

If at this point the table and handwheel dial are both at "whole-angle" positions (Figures 8-11 and 8-12, shown earlier), the setup is complete.

If not, depending on the table design, it may be possible simply to modify the angle indicators. If your table has an adjustable main-angle pointer (Figure 8-17), move it to coincide with a graduation on the table rim. This is the *reference*, or starting value. In addition, one of the whole-angle datum lines on the handwheel (0, 1, 2, or 3) has to coincide with the vernier zero (it doesn't matter which datum you choose). Loosen the set screw securing the handwheel dial, rotate the dial, and retighten. *The dial graduations and the main-angle indication on the table rim are now synchronized, provided that all table motion thereafter is in your chosen direction.*

FIGURE 8-17 Adjustable main angle pointer.

If you make a note of both the *reference main angle* and its related *datum number* on the handwheel, the starting point of the job is recoverable.

If your main-angle pointer is fixed, as in Figure 8-11, consider replacing it with a shop-made adjustable one. This can save a lot of fine-tuning effort.

Working with a fixed main-angle pointer is not so easy. Start by aligning the workpiece with a dial indicator, as above; then rotate the table *in your chosen direction* to bring the next line on the rim to the main-angle pointer.

Next, loosen the handwheel dial to set a whole-number datum against the vernier zero. Clamp the table firmly. Slightly loosen the workpiece clamps; then gently tap the workpiece around into alignment—time-consuming, but it can be done.

8-12 DRILLING THE HOLE PATTERN

For simplicity, the following steps assume that the reference main angle and the handwheel datum are both zero. In practice, they will be the reference numbers established in the setup procedure described above.

To drill the hole pattern:

1. If you have moved it, return the mill table to ABS $X = 0$, $Y = 0$ (this realigns the mill spindle with the rotary table's center of rotation).

    Leave the X axis locked. Bring the workpiece forward (Y axis) by an amount equal to the hole circle radius, R, shown in Figure 8-14.

    This places hole 1 under the spindle. Lock both the X and Y axes. From this point on, *do not touch* the X and Y hand-wheels until the job is done!

2. Clamp the rotary table at the hole 1 position; then drill the hole.

3. Unclamp the rotary table, then turn it through *exactly* 51 degrees, going very, very slowly as you approach the target value. Stop when the main angle reads 51 plus the reference value (hole #1 angle, the starting point), with the correct handwheel datum number (0, 1, 2, or 3) in line with the vernier zero (A in Figure 8-18). Referring to Table 8-1, the rotary table now needs to be turned an additional 25 minutes and 44 seconds. (If you overshoot when setting the angle, reverse the handwheel one turn, then try again.)

TABLE 8-1 Seven-hole circle divisions

Hole	Degrees	Minutes	Seconds	Closest 20" Multiple
1	0	0	0	0
2	51	25	44	40
3	102	51	29	20
4	154	17	13	20
5	205	42	58	60
6	257	8	42	40
7	308	34	26	20

Table notes: 360/7 = 51.429°. For 0.429 in minutes of arc: 0.429 x 60 = 25.74 minutes. For 0.74 minutes in seconds of arc: 0.74 x 60 = 44.4 seconds. If hole 1 is exactly at 0°, the table setting for hole 2 is 51°25'44", rounded to 51°25'40".

4.   Add 25 minutes-of-arc by rotating the handwheel 12-1/2 divisions (see B and C in Figure 8-18). If you don't need to aim for seconds-of-arc precision, simply round up the adder to 26 minutes, 13 divisions. This would place hole 2 accurately enough for most purposes, but if further precision is called for, read on.

5.   To take account of the 44 second adder for hole 2 (see Table 8-1), rotate the handwheel an additional *very small amount* for vernier coincidence at 40", the nearest multiple of 20" to 44" available with this vernier (see D in Figure 8-18). Clamp the rotary table, then drill hole 2.

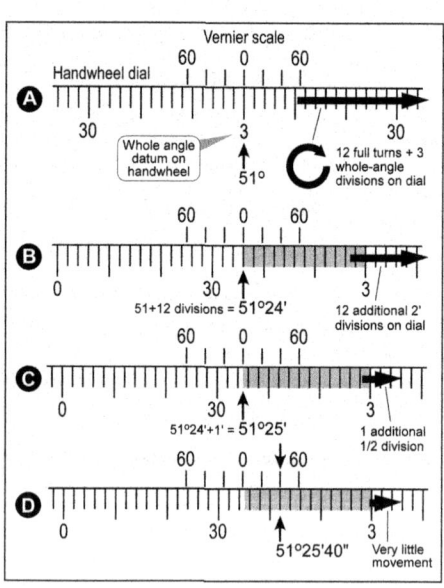

FIGURE 8-18 Positioning the table for hole 2. This diagram assumes that the reference angle is zero, and that the handwheel datum for hole 1 was also zero. In practice, these values will be different, depending on the alignment process.

6. Complete the hole circle in the same way. Each hole position is defined by the combination of a main-angle indication and handwheel datum, followed by minor adjustments to add minute and seconds-of-arc, as required. Minutes can be set directly using the handwheel dial, Figure 8-12. For "rounded" seconds, see the last column in Table 8-1.

8-13 WORK HOLDING

Users tend to accumulate over time a collection of hold-down devices, some off the shelf, mostly custom made. Some examples are shown in the following photos.

The T-slots on the rotary table look to be capable of holding anything, but that's wishful thinking. Many more hold-down options are available with an auxiliary plate, as shown in Figure 8-19. This shop-made plate is secured to the table with four flat-head screws and T-nuts.

FIGURE 8-19 Auxiliary plate. This example is steel, 8" square x 3/8" thick, with a matrix of M6 threaded holes at 1" intervals. The plate is fastened to the table by four flat-head screws and T-nuts, inset (about 0.7" x 0.25", tapped M8 or 5/16). Radius the inside edges of the T-nuts to allow placement close to the table's center.

To allow table positioning as shown in Figure 8-9, the auxiliary plate has a 3/4" hole in the center. This allows it to be used with adapter D (Figure 8-9).

In Figure 8-20, a toolmaker's vise is mounted on the steel plate with shop-made slotted clamps. The same clamps are used in Figure 8-21 to hold down an aluminum workpiece at a specific location, in this case a predrilled hole that has to be precisely aligned with the mill spindle. This is done using a 60° centering tool in a 1/2" collet (adapter B, Figure 8-9 ; this is the same arrangement used in Figure 8-15).

Assuming that you will want to: (1) center on a hole location as shown, and also; (2) have one of the workpiece edges at a specific angle, start by turning the table to a memorable angle such as 0°, 90°, 180°, or 270°; then lock the table. Now, with the workpiece firmly centered with the centering tool, spin the workpiece around to bring its reference edge as closely in line as you can with the mill's X axis. (One way to do this is to set the heel of an adjustable square against the ground front surface of the mill table; then feed the blade out to touch the workpiece, aiming for equal extension first at one end, then the other.) Fully tighten the workpiece clamps. If necessary, fine tune the setup using a dial indicator as described in Section 8-11.

FIGURE 8-20 Auxiliary plate with vise. The vise is attached by socket head screws and L-shaped clamps, shop-made from angle iron.

FIGURE 8-21 Centering a workpiece.

Figure 8-22 shows one way to attach an independent 4-jaw chuck to the rotary table. This is a 4" chuck removed from its lathe backplate. The substitute backplate here is a 3/4"-thick PVC disc with a 1/4" center hole. This allows it to be precisely located on the table using adapter C in Figure 8-9. The inset photo in Figure 8-22 shows a 1/8"-thick center disc exactly matching the chuck's locating bore—not strictly necessary, but good for preserving the axial centerline when transferring a workpiece in the chuck from lathe to mill.

FIGURE 8-22   PVC backplate for lathe chuck. The material used is 3/4"-thick PVC—easily machinable, rock solid, and much less expensive than metal alternatives.

8-14 AN EASIER WAY TO DIVIDE THE CIRCLE

We have seen how the rotary table in its standard form makes angular positioning possible in degrees, minutes, and seconds of arc, with sufficient accuracy for any purpose in the small shop. However, there's no denying it can be tedious.

For example, suppose you need a pattern of 50 equally spaced holes, angular spacing 360/50 = 7.2° = 7°12' (7°, 12 minutes of arc). You would drill the first hole at 0°, the second hole at 7°12', the third hole at 14°24', the fourth hole at 21°36", and so on. There is nothing difficult about this, but it would call for a spreadsheet of the 50 angles, plus the need to inch up carefully to each value in turn, always in the same direction, exactly as described in Section 8-12. You could label this the "analog division" method.

With a dividing plate add-on, there is a much easier method, "digital division" (provided the hole count is one of those listed in the Tables 8-3 and 8-4 later in this chapter). The dividing plates are predrilled with precise locating holes, 18-hole circles in total, so there are no approximations and no guesswork.

The three dividing plates in the set (Figures 8-23 and 8-28) cover 72 hole counts (or number of gear teeth) in the range 1 to 90, plus many more between 91 and 360. Hole or gear tooth counts are known as N values. (For divisors not available with the dividing plates, set the table by degrees, minutes, and seconds in the ordinary "analog" way.)

Each of the three dividing plates has six concentric circles of equally spaced holes, which prompts this question: *How can 18-hole circles deliver four times that number of N values? They* can, because the system uses a combination of hole counting and a specific *number of rotations* of the worm shaft. Dividing plates come with an assortment of ancillary hardware, as shown in Figure 8-24.

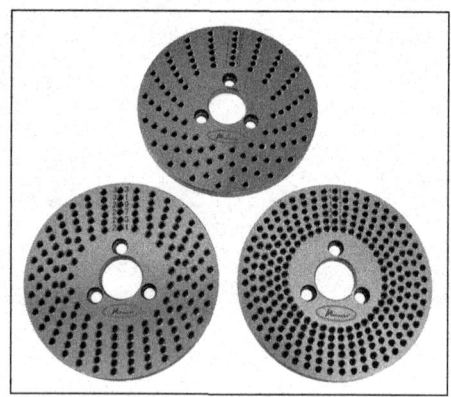

FIGURE 8-23 Dividing plates for the rotary table.

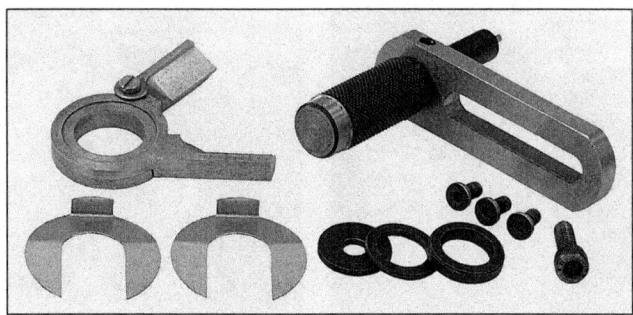

FIGURE 8-24 Dividing plate hardware. *Left:* Sector arm assembly (brass). *Right:* Crank arm assembly, with spring-loaded indexing pin. *Bottom left:* C-shaped springs to retain the sector arm assembly.

8-15 INSTALLING A DIVIDING PLATE

There are minor differences among dividing plate sets. The following applies only in general terms.

Before removing the handwheel, check that the table has been adjusted for smooth rotation with minimum backlash.

To install a dividing plate:

1. Remove the handwheel and graduated dial from the worm shaft. Set them aside for the next time you want to operate the table in the ordinary way (don't lose the key).

2. Referring to Section 8-17, select one of the three dividing plates (Figure 8-25).

3. Attach the dividing plate to the clutch lever backplate (this has 60-0-60 vernier marks—see Figure 8-12).

4. Install the brass sector arm assembly on the worm shaft sleeve. Loosen the clamp screw on the assembly to check that the arms open and close freely. Leave the screw loose at this

FIGURE 8-25   Dividing plate installed.

FIGURE 8-26   Sector arm assembly.

time. (When set to a specific span, the clamp screw is tightened, and the assembly is held in place by a C-shaped thin-metal spring.)

5. Place spacer washers (one or more) in front of the sector arms, covering the threaded collar, to provide a stable surface for the crank arm (Figure 8-26).

6. Place the crank arm on the worm shaft, securing it with screw and washer (Figure 8-27). *The knurled crank handle is spring-loaded. When released, the spring presses the indexing pin against the dividing plate.*

7. While holding the indexing pin clear, slide the crank arm into position so that the pin is at the radius of the selected hole circle.

FIGURE 8-27   Crank arm installed.

8. Fully tighten the crank arm screw. While holding the spring-loaded indexing pin clear, Figure 8-27, rotate the arm a few times around the circle

to be sure that the indexing pin locates cleanly in several holes in the circle and *does not disturb the sector arms.* Very important!

8-16 USING THE DIVIDING SYSTEM

Key fact:

The drive ratio between the worm shaft and table is 90:1, meaning that it takes 90 revolutions of the crank arm to rotate the table 360°. Put another way, one complete revolution of the handwheel turns the table 4°.

It follows that any angular spacing that is a multiple of 4° can be set simply by counting turns of the crank arm, starting and ending at a specific hole on any dividing plate. (You don't even need a dividing plate to do this. Use the handwheel instead; just be sure you begin and end at the same whole-angle datum on the handwheel dial—0, 1, 2, or 3.)

In these notes, N, the divisor, is the number of machining events around the circle (gear teeth to be cut, holes to be drilled, etc.). Angular spacing for machining purposes is 360° divided by N.

Example: For a 30-tooth gear, the angular spacing is 360/30 = 12°, calling for three revolutions of the crank arm.

Other "easy" values of N are given in Table 8-2.

TABLE 8-2   Easy values of N

N	2	3	5	6	9	10
Turns	45	30	18	15	10	9
N	15	18	30	45	90	
Turns	6	5	3	2	1	

8-17 WHICH DIVIDING PLATE?

This is given by the formula 90/N. The result is a whole number, which is the *number of crank turns,* plus a *fractional remainder* that indicates which dividing plate to use (choice of three plates, Figure 8-28).

Example: For a 38-tooth gear, N = 38, 90/38 = 2 + 14/38. This means **2** full turns of the crank, plus **14** intervals (spaces) on a 38-hole circle. There is no 38-hole circle, but the **14/38** fraction can be converted to **7/19**, exactly the same value.

Now we are in business, because disc A (Figure 8-28) has a 19-hole circle.

Generally speaking, you have a usable solution if multiplying or dividing the fractional remainder by a whole number gives a denominator (19 in the above case) corresponding to one of the 18-hole circles. For the correct number of spaces to step around the hole circle (seven in the above case), be sure to apply the *same multiply or divide factor to both numbers* of the fraction.

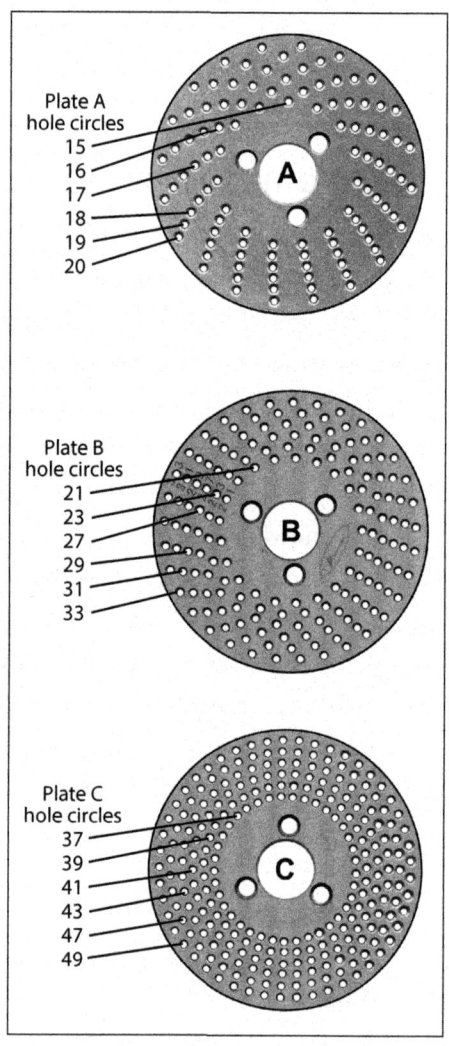

FIGURE 8-28 Dividing plate detail.

8-18 DRILLING
A 27-HOLE CIRCLE

Table 8-3 (Section 8-19) and Table 8-4 (Section 8-20) list all possible values of N from 2 to 360 with the three dividing plates usually supplied. Note that in some cases there is a choice of plate and hole count; e.g., for N = 27 there are six choices using all three plates. The 27-hole illustration described here uses plate B, hole circle 21 (Figure 8-28). The corresponding entry in Table 8-3 reads "3+7/21," meaning that the job calls for a series of three full turns of the crank plus seven additional spaces.

The process consists of eight steps, shown in Figures 8-29 to 8-31.

1. Start the process by turning the crank arm clockwise to bring the table to a "memorable angle" such as 0°, 90°, 180°, or 270° on the main-angle indicator.

2. Set the sector arms to span seven *spaces* on the 21-hole circle of the dividing plate; then tighten the sector arm clamp screw.

3. Set the crank arm radius so that the *indexing pin is on the 21-hole circle.* Insert the indexing pin in your choice of starting location.

4. Clamp the table; then drill hole 1.

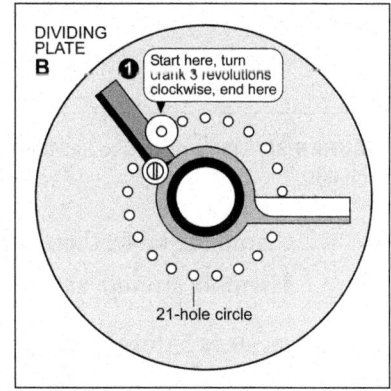

FIGURE 8-29 Preparation for hole 2: crank arm position after three turns from the hole 1 location.

5. Swing the sector arm assembly around to abut the pin, as in Figure 8-29. Unclamp the table.

When cranking, be sure to keep a grip on the spring-loaded indexing pin. If you let it slip, it can accidentally swipe the sector arms—possibly a fatal, start-over error.

6. Pull back the knurled handle; then crank the arm three full turns clockwise (12° of table rotation), stopping at the same hole location as at the start (Figure 8-29).

7. Continue turning the crank to bring the indexing pin into contact with the opposing-sector arm (Figure 8-30).

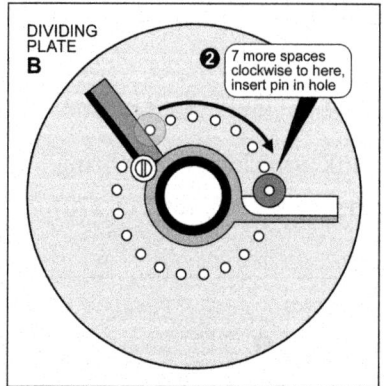

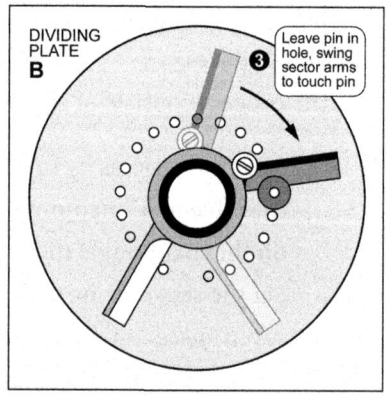

FIGURE 8-30 Table moved to location for hole 2.

FIGURE 8-31 Sector arms positioned for the next move.

8. Clamp the table; then drill hole 2. Swing the sector arm assembly around, as in Figure 8-31.

Repeat steps 5 through 8 to complete the 27-hole project.

8-19 DIVISIONS FROM N = 2 TO N = 90

In Table 8-3 most N values, aside from the handful that can use any plate, call for a whole number of crank turns, plus a fractional turn. For example: when machining a 33-tooth gear, each cutting pass is followed by two full turns of the crank, plus 24/33 of a turn, indexed using plate B. The underlying math here is straightforward:

The angular spacing for 33 teeth is 360/33 = 10.90909°. Two full turns of the crank rotate the workpiece 8°, leaving a balance of 2.90909°. That is exactly the same as the fractional turn 24/33, when multiplied by 4:

$$4° \times 24 \div 33 = 2.90909°$$

Several values of N can be indexed in two or more ways. Two examples: N = 14, N = 27. This can sometimes eliminate the need to swap plates when doing a series of dividing plate jobs. Values of N that can be indexed simply by turning the crank arm are shaded gray: Values not indexable with dividing plates are marked N/A.

TABLE 8-3   Values from N = 2 to N = 90

N (divisions)	90/N (turns + spaces)	Plate	N (divisions)	90/N (turns + spaces)	Plate
2	45	Any	26	3 + 18/39	C-39
3	30	Any	27	3 + 5/15	A-15
4	22 + 10/20	A-20	27	3 + 6/18	A-18
4	22 + 9/18	A-18	27	3 + 7/21	B-21
4	22 + 8/16	A-16	27	3 + 9/27	B-27
5	18	Any	27	3 + 11/33	B-33
6	15	Any	27	3 + 13/39	C-39
7	12 + 18/21	B-21	28	N/A	
7	12 + 42/49	C-49	29	3 + 3/29	B-29
8	11 + 5/20	A-20	30	3	Any
8	11 + 4/16	A-16	31	2 + 28/31	B-31
9	10	Any	32	2 + 13/16	A-16
10	9	Any	33	2 + 24/33	B-33
11	8 + 6/33	B-33	34	2 + 11/17	A-17
12	7 + 8/16	A-16	35	2 + 12/21	B-21
13	6 + 36/39	C-39	35	2 + 28/49	C-49
14	6 + 9/21	B-21	36	2 + 8/16	A-16
14	6 + 21/49	C-49	37	2 + 16/37	C-37
15	6	Any	38	2 + 7/19	A-19
16	5 + 10/16	A-16	39	2 + 12/39	C-39
17	5 + 5/17	A-17	40	2 + 4/16	A-16
18	5	Any	41	2 + 8/41	C-41
19	4 + 14/19	A-19	42	2 + 3/21	B-21
20	4 + 8/16	A-16	42	2 + 7/49	C-49
21	4 + 6/21	B-21	43	2 + 4/43	C-43
21	4 + 14/49	C-49	44	N/A	
22	4 + 3/33	B-33	45	2	Any
23	3 + 21/23	B-23	46	1 + 22/23	B-23
24	3 + 12/16	A-16	47	1 + 43/47	C-47
24	3 + 15/20	A-20	48	1 + 14/16	A-16
25	3 + 9/15	A-15	49	1 + 41/49	C-49
25	3 + 12/20	A-20	50	1 + 12/15	A-15

(continued on next page)

TABLE 8-3   Values from N = 2 to N = 90 (continued)

N (divisions)	90/N (turns + spaces)	Plate	N (divisions)	90/N (turns + spaces)	Plate
50	1 + 16/20	A-20	69	1 + 7/23	B-23
51	1 + 13/17	A-17	70	1 + 6/21	B-21
52	N/A		71	N/A	
53	N/A		72	1 + 4/16	A-16
54	1 + 10/15	A-15	73	N/A	
54	1 + 12/18	A-18	74	1 + 8/37	C-37
54	1 + 14/21	B-21	75	1 +3/15	A-15
54	1 + 18/27	B-27	75	1 + 4/20	A-20
54	1 + 22/33	B-33	76	N/A	
54	1 + 26/39	C-39	77	N/A	
55	1 + 21/33	B-33	78	1 + 6/39	C-39
56	N/A		79	N/A	
57	1 + 11/19	A-19	80	1 + 2/16	A-16
58	1 + 16/29	B-29	81	1 + 2/18	A-18
59	N/A		81	1 + 3/27	B-27
60	1 + 8/16	A-16	82	1 + 4/41	C-41
61	N/A		83	N/A	
62	1 + 14/31	B-31	84	N/A	
63	1 + 9/21	B-21	85	1 + 1/17	A-17
64	N/A		86	1 + 2/43	C-43
65	1 + 15/39	C-39	87	1 + 1/29	B-29
66	1 + 12/33	B-33	88	N/A	
67	N/A		89	N/A	
68	N/A		90	1	Any

Alternates missing from Table 8-3

N (divisions)	90/N (turns + spaces)	Plate	N (divisions)	90/N (turns + spaces)	Plate
12	7 + 9/18	A-18	40	2 + 5/20	A-20
12	7 + 10/20	A-20	60	1 + 10/20	A-20
20	4 + 9/18	A-18	63	1 + 21/49	C-49
20	4 + 10/20	A-20	70	1 + 14/49	C-49
36	2 + 9/18	A-18	72	1 + 5/20	A-20
36	2 + 10/20	A-20			

8-20 DIVISIONS FROM N = 91 TO N = 360

TABLE 8-4 Values from N = 91 to N = 360

N (divisions)	90/N (turns + spaces)	Plate	N (divisions)	90/N (turns + spaces)	Plate
93	30/31	B-31	162	10/18	A-18
94	45/47	C-47	165	18/33	B-33
95	18/19	A-19	170	9/17	A-17
96	15/16	A-16	171	10/19	A-19
98	45/49	C-49	174	15/29	B-29
99	30/33	B-33	180	8/16	A-16
100	18/20	A-20	185	18/37	C-37
102	15/17	A-17	186	15/31	B-31
105	18/21	B-21	189	10/21	B-21
108	15/18	A-18	190	9/19	A-19
110	27/33	B-33	195	18/39	C-39
111	30/37	C-37	198	15/33	B-33
114	15/19	A-19	200	9/20	A-20
115	18/23	B-23	205	18/41	C-41
117	30/39	C-39	207	10/23	B-23
120	12/16	A-16	210	9/21	B-21
123	30/41	C-41	222	15/37	C-37
126	15/21	B-21	225	8/20	A-20
129	30/43	C-43	230	9/23	B-23
130	27/39	C-39	234	15/39	C-39
135	10/15	A-15	235	18/47	A-18
138	15/23	B-23	240	6/16	A-16
141	30/47	C-47	243	10/27	B-27
144	10/16	A-16	245	18/49	C-49
145	18/29	B-29	246	15/41	C-41
147	30/49	C-49	255	6/17	A-17
150	9/15	A-15	258	15/43	C-43
153	10/17	A-17	261	10/29	B-29
155	18/31	B-31	270	6/18	A-18
160	9/16	A-16	279	10/31	B-31

(continued on next page)

TABLE 8-4   Values from N = 91 to N = 360 *(continued)*

N (divisions)	90/N (turns + spaces)	Plate	N (divisions)	90/N (turns + spaces)	Plate
282	15/47	C-47	315	6/21	B-21
285	6/19	A-19	324	5/18	A-18
288	5/16	A-16	330	9/33	B-33
290	9/29	B-29	333	10/37	C-37
294	15/49	C-49	342	5/19	A-19
297	10/33	B-33	345	6/23	B-23
300	6/20	A-20	351	10/39	C-39
306	5/17	A-17	360	4/16	A-16
310	9/31	B-31			

8-21 USING THE TAILSTOCK

The tailstock option provides support for long workpieces with the rotary table mounted vertically (Figure 8-32). In this example, the workpiece is center-drilled at the tailstock end and is held at the other end in the 3-jaw chuck usually available as an option for the rotary table.

Alternatively, if an MT2 dead center is available, the workpiece could be center-drilled at both ends, and then held between centers, as on a lathe (in this case the rotary table end of the workpiece would need to be anchored to a driving device similar to a lathe dog).

The tailstock must first be set up so that its center is at the same height as the rotary table's center of rotation. This is a one-time operation. If an MT2 dead center is available to fit the center hole, see Figure 8-33 for a quick way to set height with reasonable accuracy.

FIGURE 8-32 Vertically mounted rotary table with tailstock.

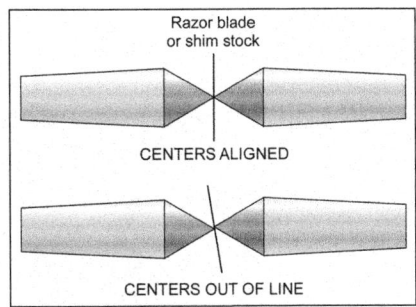

FIGURE 8-33 Quick check for center-to-center alignment.

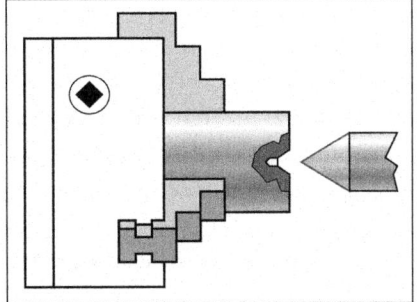

FIGURE 8-34 Tailstock height set by center-drilled rod.

For more precise alignment, center-drill a short length of round rod with minimal TIR using (recommended) a 4-jaw chuck in the lathe. Hold the bar in the 3 jaw chuck; then locate the tailstock center in the center-drilled cavity (Figure 8-34).

Alignment of the tailstock relative to the rotary table in the horizontal plane (the mill table) is more complicated, because there are no machined reference surfaces to use as a guide. As a starting point, set the heel of a precision square against the front surface of the mill table; then slide the vertical surface of the rotary table up to the blade (Figure 8-2 shows how this is done, but with the rotary table lying flat). Position the rotary table symmetrically on the mill table so that its centerline lies directly over the middle T-slot. Centerline positioning like this is important, because the tailstock has only a limited amount of front-to-back movement when its hold-down screws are installed.

Front-to-back alignment of the tailstock can now be checked using a setup as shown in Figure 8-32. In this illustration a dial indicator is held in a collet (or drill chuck). The objective here is to minimize indicator movement as the table is run back and forth—at the end of each pass, gently tap the tailstock to the front or back until the deviation is acceptable.

Q & A

Q: I can use my spin indexer for gear cutting. Do I need a rotary table as well?

A: Maybe not, but you will certainly need to use something other than a spin indexer if the result of dividing 360 by the number of gear teeth to be cut is not a whole number (60 teeth, 6° spaces, OK; 16 teeth, 22.5° spaces, not OK).

CHAPTER 9

Installing Two Popular Accessories

CONTENTS AT A GLANCE

9-1 ABOUT THIS CHAPTER

This chapter is an overview of the hardware issues involved in adding *DRO scales* to a milling machine, and adding *a power feed unit* to motorize left-right motion of the table. The information given here is not specific to any one product on the market (there are many); what is intended is a brief description of some of the issues you may encounter when doing your own installation—as opposed to buying preinstalled.

9-2 OPTICAL DRO SCALES

Position reporting systems (linear encoders) for machine tools have been around for more than 60 years. The original technology was based on detecting "moiré fringe" patterns that occur when two optical grids (parallel lines) are superimposed, one at a slight angle to the other (Figure 9-1).

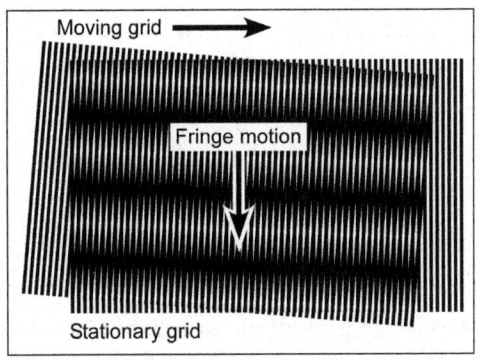

FIGURE 9-1 Moiré fringes caused by optical interference between two grids.

In Figure 9-1 the moving grid is inclined at 5°. The dark fringes move up or down, depending on left-right motion of the moving grid. The fringe motion is more than 10 times the amount of the moving grid's motion, a "zoom factor" that is the key to all moiré-based systems—a large, easily detectable optical effect caused by a tiny physical movement.

Early versions of the grids were photo images of alternating black and transparent lines. Photo images are still used in some applications, but optical scales today are usually finely etched glass substrates. The moving element of the "grid pair" is housed in a *reading head*, 2" or 3" long, that runs along the stationary grid, much like a tram on rails (Figure 9-2).

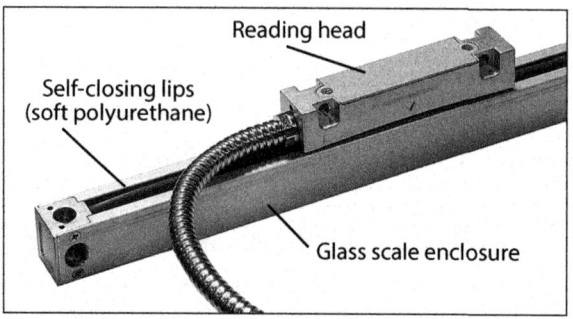

FIGURE 9-2 Typical glass scale.

The stationary grid has to be slightly longer than the maximum travel of the axis to be measured—table, cross-slide, or headstock/knee—*plus* the length of the reading head.

Glass grids (aka scales) are precise, predictable, and very reliable. There is the obvious problem of mechanical fragility; however, if the scales are properly installed, they are out of harm's way, protected by covers to keep out chips and coolant. The other issue with glass scales is the need for precise alignment of reading head and scale, to within very few mils from one end to the other. This should not deter you from installing your own scales, but do plan on a lengthy work session with some fine tuning.

9-3 MAGNETIC DRO SCALES

An interesting alternative to glass scales has been on the market for the past 15 years. This is the magnetic DRO scale, one example of which is shown in Figure 9-3.

Magnetic scales like the one in the figure are capable of the same resolution as the typical glass scale, namely one micron (approximately 0.00004"). They have several advantages:

- **Robustness.** A magnetic scale intended for a machine tool is typically an aluminum extrusion containing a precisely magnetized barium ferrite tape, protected by a thin stainless-steel cover. The reading head is fully encapsulated and therefore

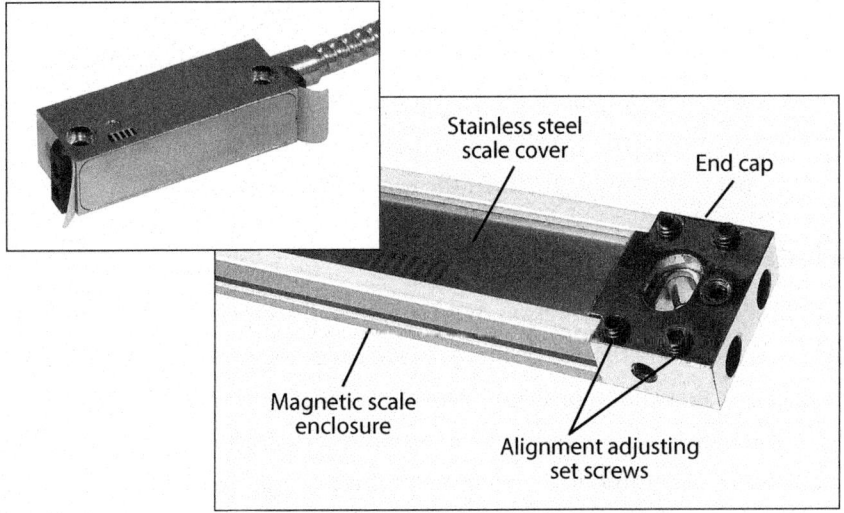

FIGURE 9-3 Typical magnetic scale. The reading head (see the inset, upper left) rides along the surface of the scale cover, separated (in this example) by a gap of 0.02".

resistant to contaminants—but that doesn't eliminate the need for a cover.

- **Sizability.** Magnetic scales can usually be cut to length in the shop to fit exactly the available space on the machine (a fine-tooth hacksaw is better for this than a cut-off blade, which can burn the magnetic material). It is usually a simple job to remove one of the end caps and reattach it to the shortened extrusion.

- **Compactness.** The cross section of a magnetic scale is typically less than that of a glass scale intended for the same application.

- **Easier installation.** Magnetic scales are usually self-support-ing, meaning they may not need an intermediate support beam—a space-saving bonus. Also, the scale end caps typi-cally come with built-in set screws to adjust separation and tilt relative to the machine casting.

- **Less critical alignment.** Having recently installed eight glass scales and six magnetic scales, I am reasonably certain that the magnetic scales are the more forgiving—certainly easier to handle and less susceptible to mechanical damage. The manufacturers' data sheets for the two scale types were equally stern regarding installation do's and don'ts, but my bet is still on magnetic as the easier way to go.

9-4 GENERAL NOTES

This first note is an observation that applies to all scale types. They are available in various lengths, so the first job is to order the exact sizes needed. An overlong scale is not a problem, provided there is room enough to install it (or if magnetic, can it be cut to length?). A too-short scale, meaning one that cannot accommodate end-to-end travel of the axis in question, is a real limitation. Exchange it for one of the right length if you can, or—worst case—install it with external mechanical stops to prevent reading head collisions and worse.

Also, before ordering, be sure that there is suitable space available. Test-fit lengths of wood in your chosen X, Y, and Z axis locations. Look out for possible interactions between the DRO scale/reading head and clamp screws, handles, travel stops, etc.

Another observation, this one seeming to apply to all scale types, is that they come with a selection of brackets and other hardware, usually without assembly instructions. What this amounts to is that you are on your own. The hardware supplied with the scales is rarely enough to complete the installation, so be prepared to add your own materials, as in Figures 9-5 and 9-6, shown later in the chapter. For those two installation examples it would have been possible to use, say, 50% of the supplied brackets, but the result would have been uglier, and probably less robust. (Another factor was the easy availability of aluminum extrusion from a local hardware store.)

9-5 DRILLING AND TAPPING CAST IRON

This can be a nail-biting experience, but all it takes are a steady hand and a hand drill with a true-running chuck. It may also be helpful to have a set of transfer punches—insert them through holes in the DRO components to mark drilling locations on the mill (alternatively, mark with a fiber-tip pen; then use an automatic center punch). Cast iron can be machined dry (though some machinists prefer a light lubricant/solvent like WD-40). Most items are attached using either #8 and #10 socket head cap screws or, if you want to keep the entire machine metric, M4 and M5.

It is important to keep the *drill and tap at right angles* to the machine axes. You can buy drill and tap guides, but they will likely not work in the desired locations. Custom guides can be made in the shop from 1/2"- or 3/4"-thick material, shaped if necessary to sit squarely on the curved/angled machine surface, gripped only by the fingers (or align the drill bit using a mini-size spirit level). Use stub-length drills if you have them. To reduce the force needed to tap the hole, consider using a drill *slightly larger* than the standard size given in the charts. If using longer bits, use the absolute minimum pressure needed to drill—excess pressure will bend the drill, flaring the hole. Cast-iron dust gets everywhere, so mask off the surrounding area.

According to some sources, 3-flute taps are less prone to breakage than 2-flute or 4-flute taps, but all of them are fragile in #8 and #10 sizes (or M4 and M5). Some machinists start the thread with a *plug taper* tap and then replace it with a *bottoming taper* to finish the thread. Whatever else you do, align the tap(s) as carefully as you drilled the holes. Taps break easily, so take your time—small forward rotations, frequent reverses to clear the chips. Some installers run the taps in with a hand drill that's set at low speed, but that calls for a degree of confidence most of us don't have. Use a pipe cleaner, and vacuum out when finished.

Brand-name taps cost a lot more than generic types, but they may be worth it in this instance. If you do break a tap, usually nothing can be done other than to drill and tap in another location alongside. With luck, only you will know. (Tap removers sound tempting, but success with these small sizes is a long shot.)

9-6 INSTALLING THE SCALES AND READING HEADS

This is the stage in the process that sometimes calls for equal parts of ingenuity and patience.

The scales have to be installed *precisely* in line with the axis motion, a problem in most cases because the painted cast-iron surfaces are anything but flat and are not aligned with the lead screws.

9-7 X AXIS SCALE

Fortunately, there is no curvature on the machined back surface of the table, the usual location for the X axis scale. This is the one scale that typically doesn't need an intermediate support beam—simply attach the scale, glass or magnetic, to the table with two socket head cap screws, one at each end. All scales have oval-shaped attachment holes to permit, say, 3/16" vertical adjustment.

If your X axis scale came with a narrow intermediate beam that will fit neatly at the back of the table, there is no reason not to use it, other than the fact that it will reduce Y axis travel by the beam's thickness—not usually an issue. One reason to use the intermediate beam is easier installation of the scale cover; the beam will likely have been drilled and tapped for two cover screws, one at each end. Otherwise, you will need to attach the cover with

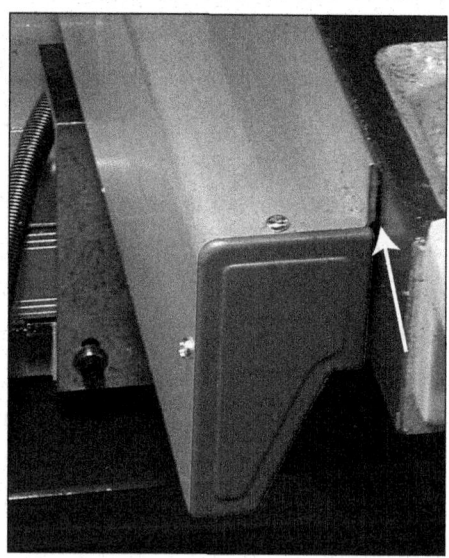

FIGURE 9-4 An X axis scale cover attached with heavy-duty adhesive tape, indicated by the arrow.

screws in two additional drilled and tapped holes. A less time-consuming solution is to use double-sided adhesive tape such as 3M Super Strength

Molding Tape. The installation in Figure 9-4 has survived years of oils and coolant splashes.

Glass X axis scales are installed *upside down*, with the reading head attached to the back of the mill saddle—likely a painted, unmachined surface, so expect to spend a lot of time shimming the reading head for accurate alignment with the scale.

9-8 Y AXIS SCALE

The Y axis scale is installed on a painted, non-flattened surfaces and will need careful alignment *before* you spend time on its reading head. (This also applies to the Z axis scale, see Section 9-9.) Figure 9-5 shows a Y axis installation using a support beam that came with tilt-adjusting set screws. (You can see here that the upper screw pushes the beam outward to correct for the slope on the main casting.) The beam in this case, as supplied, was drilled and tapped for the scale attachment screws. Installation of the beam was easy—drilling and tapping for two M6 screws, one at each end. Vertical alignment of the beam is not critical, because the scale attachment allows considerable up-and-down adjustment.

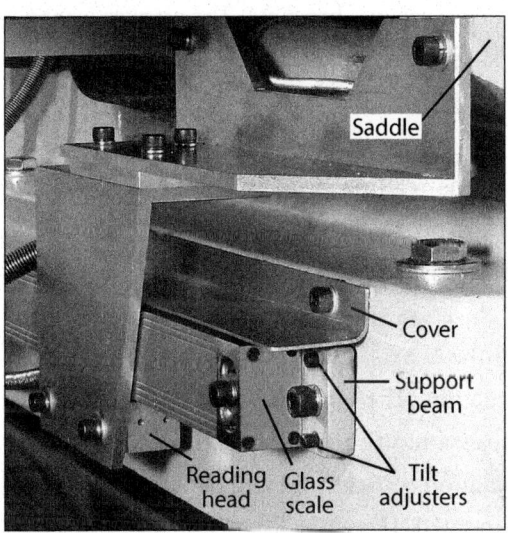

FIGURE 9-5 Custom-made brackets for a Y axis installation.

The more complicated part of the job was constructing the coupling pieces from aluminum extrusion. Even with careful "right-angles-every-where" construction, expect to spend considerable time shimming the reading head for precise alignment. It may be helpful to check the tram of the scale (horizontal and vertical) using a dial indicator clipped to the reading head bracket.

None of this construction would have been called for if the hardware that came with the scale had been adequate. This may be an extreme case, but the word from mechanics who install these scales every day is that there is always a need for a lot of adaptation. For instance, the intermediate support beams usually supplied with glass scales often have no means of adjustment. In such cases, drill/tap the support beams for set screws on the lines of Figure 9-5.

The scale cover in Figure 9-5 was an afterthought, the theory being "no need," because the reading head is protected by the scale itself, which also is shielded by the table much of the time. Correct on both counts, but it was clear from the first use that the cover was not optional. This could have been the conventional cover, as in Figure 9-4, but that would have called for remanufacturing various brackets. The answer was to remove the riser panel from the supplied cover.

9-9 Z AXIS SCALE

Coupling the Z axis reading head to the headstock or knee is similarly complex, but easier because the scale is out in the open. Figure 9-6 shows one example. The scale can usually go on either side, unless prevented (as in this case) by elevation stops, clamps, etc.

Just as with the Y axis, the supplied hardware was not helpful, hence the DIY construction. The good news in this case—no need for a cover (coolant splashes shouldn't get that far). Even better news is the fact that this Z axis installation is rock solid and can be used reliably for machining to a depth within ± 0.001".

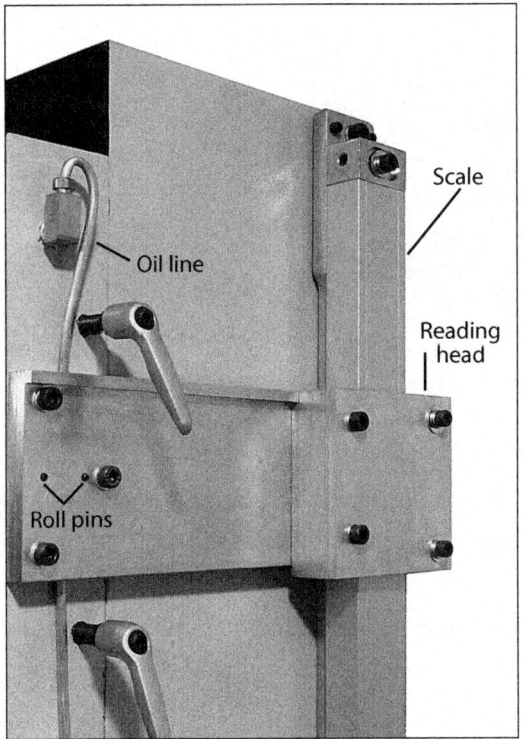

FIGURE 9-6 Z axis installation. The roll pins were added to ensure stability even with an overall bracket length of about 10".

9-10 INSTALLING A TABLE POWER FEED UNIT

Compared with installing DRO scales, this is a breeze—just one axis (X), mechanical simplicity, and all necessary parts (usually) supplied with the power unit (assuming you bought it from a supplier that guaranteed it would fit your mill).

Table power feed units are all basically similar, with a left-right direction lever and a speed control potentiometer. Typically, there is also a push button labeled "RAPID" or "FAST TRAVERSE," which drives the table at maximum speed (regardless of the potentiometer setting) in the direction selected by the lever (Figure 9-7).

FIGURE 9-7   Typical table power feed unit.

To allow full access to the Y axis clamps, bench mill power feed units are usually installed at the left-hand end of the table. (Knee mill power feeds vary—some are custom made for specific mills.) Start the installation by removing the crank handle/handwheel and graduated collar from the left-hand end of the table lead screw. These items will not be needed and can be set aside.

The left-hand end of the lead screw may need to be extended to allow proper positioning of the driven gear. If so, an extender will be included in the kit. Install both the extender and driven gear, tightening both set screws (Figure 9-8).

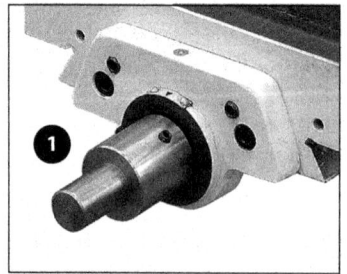
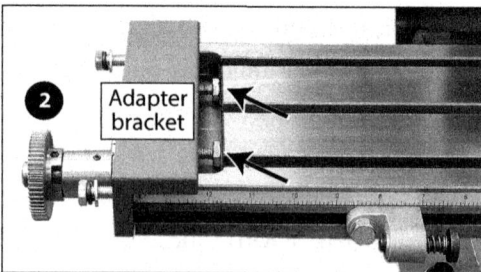

FIGURE 9-8   Extender (1) and driven gear (2). The adapter bracket in this example is clamped to the table casting by hex head screws indicated by the arrows.

Depending on the mill–power unit combination, there are two ways of attaching the power feed unit. One method calls for an adapter bracket that hooks onto the well at the end of the table, as in the photo on the right in Figure 9-8. *Avoid overtightening the screws (indicated by the arrows)—this can split the adapter bracket.*

FIGURE 9-9   Hole locations for direct attachment.

In the second method the power feed unit is attached directly to the table casting with two screws in drilled and tapped holes (Figure 9-9). *Take care with this—there is little room for error.*

The next step is measuring the distance from the outer face of the adapter bracket (Figure 9-8, *right*) or from the end of the table (Figure 9-9).

With that number in mind, loosen the socket head cap screws securing the (usually preinstalled) motor bracket (Figure 9-11). Adjust the distance between the bracket and motor gear to match the measured distance. Use a depth gauge to ensure the motor body and bracket are perfectly square— *this is essential to minimize gear*

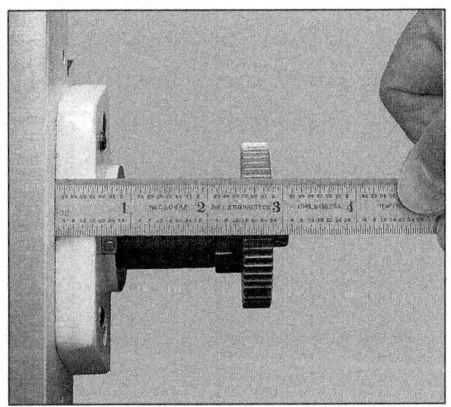

FIGURE 9-10   Measure between the driven gear and attachment surface.

noise. When assembled, make sure the driven gear is clear of the motor housing.

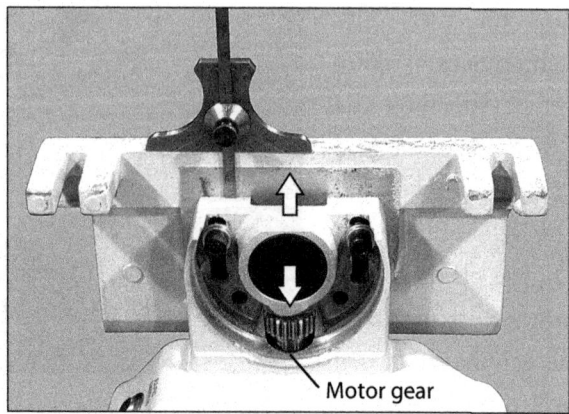

FIGURE 9-11   Adjust the motor bracket to position the motor gear at the same separation from the attachment surface as the driven gear, Figure 9-10.

9-11 GEAR ENGAGEMENT AND TRIAL RUNNING

"Proper engagement" of the gears is hard to specify. The separation that should exist between the gears can be gauged using a strip of standard printer paper, usually about 0.004" thick. Lower the motor unit to fully engage the gears, back the gears, back it off a few mils, and then lightly snug the attachment screws. Insert the paper strip between the gears; then crank the right-hand table handwheel. Lower the motor unit to the point where the paper feeds through with some resistance, but without binding. Grease the gears with NLGI #2, or similar.

Make sure the X axis locks are free.

Connect 110-V ac power. Set both *travel stop bumpers well off to each side*. Set the table to mid-travel. It should be free to move with the handwheel. Turn the speed control knob (potentiometer) OFF, fully counterclockwise; then set the direction lever to either left or right. Turn the speed control knob clockwise gently to start the motor. Run the table *slowly* a few inches; then set the direction lever to OFF, midway.

Always allow the motor to stop before changing direction.

Reverse the direction, and repeat the test. If gear noise is an issue, try raising or lowering the motor unit—but don't expect a totally silent drive (these are straight-cut gears). Fully tighten the screws securing the motor unit in place.

Holding the limit switch (Figure 9-12) in your hand, and with the table moving under power, test each side plunger in turn—for instance, the right-hand plunger should stop leftward motion of the table. When satisfied that the limit switch stops motion in both directions, install it in place of the solid center stop that came with the mill. Position the two travel stop bumpers a few inches to the left and right of the table center point; then try

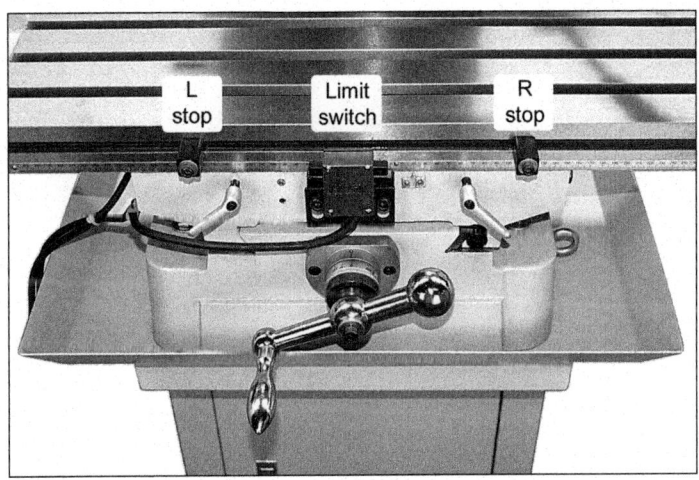

FIGURE 9-12 Limit switch.

running the table again. The limit switch may need adjustments, possibly a packing piece, so you can be sure it is actuated reliably by the bumpers.

Note: If you find the limit switch more of a hindrance than a help (as do many machinists), simply remove it from the saddle casting and allow it to dangle beneath the motor unit.

When you are satisfied that the power feed unit is fully functional, install the gear cover. This may be one of at least two types: One is a plastic molding, such as that shown in Figure 9-7, held in place by double-sided sticky pads (or drill and tap for screws). The other is a metal clip-on bracket, as shown in Figure 9-13.

FIGURE 9-13 Clip-on gear cover.

CHAPTER 10

Making a Vise for Large Workpieces

CONTENTS AT A GLANCE

10-1 THE GENERAL-PURPOSE TWO-PART VISE

Every once in a while, a job comes along that is beyond the capacity of your milling machine vise. This leaves you with two choices: Either you can spend a few hours putting together a specialized one-time clamping setup, or you can take a little more time to construct a general-purpose two-part vise (Figure 10-1). The vise described in this chapter is for a 9" x 35" mill, with an X axis capacity of about 21". The dimensions *can easily be scaled to suit any mill,* even down to the smaller machines, like the example shown in Figure 10-2. Figure 10-3 shows the main components of the vise.

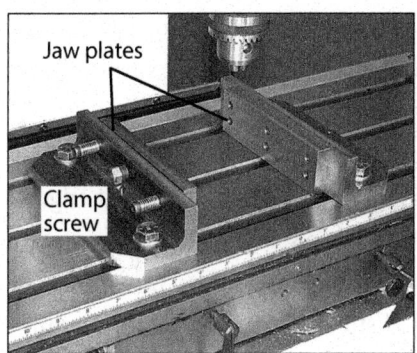

FIGURE 10-1 Complete two-part vise. The two halves are similar, except that the jaw plate on the right is fixed, but the plate on the left is moved by screw action.

FIGURE 10-2 Two-part vise in use. This is a small benchtop machine, shown here trimming the edge of a 12" x 8" x 3/8" plate.

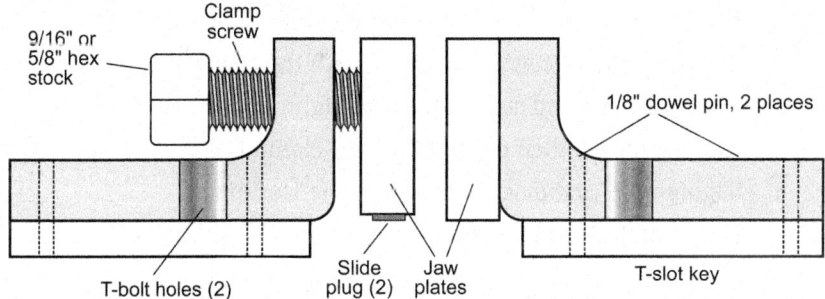

FIGURE 10-3 Vise schematic. A key part of this design is the T-slot key, which ensures that the bases are at correctly positioned on the table—just like a standard milling vise.

Cut the vise bases from 4" x 4" x 1/2" *hot-rolled structural angle iron* (see Section 10-3). You could get by with 3" x 3" instead—less metal to remove—but you would need two additional 6" x 1-3/4" x 1/2" pieces for the jaw plates (if you go with 4" x 4", you can use instead the material left over after cutting the bases).

10-2 A WORD OF CAUTION ABOUT TWO-PART VISES

The moving jaw doesn't slide on machined guides as it does in a regular mill vise, so the question is, *What gives it direction?* The answer is . . . Nothing, other than the workpiece itself and the pushing action of the clamp screw. There are two "tilt axes," vertical and horizontal. The moving jaw will tend to swing about the vertical axis if the clamped surface is out of parallel with the opposing fixed jaw, usually a minor problem.

The main concern is the horizontal axis, front to back. Here tilt depends on the relative height over the mill table of the clamped edge of the workpiece versus the centerline of the clamp screw. Other than massive construction, out of character for the small shop, there is no fix for tilt. The answer is to live with it by using parallel standoffs and the absolute minimum of clamp screw motion, like this:

1. Set the right-hand base in place, bolts tightened.
2. Back off the clamp screw for minimal separation between the moving jaw plate and the left-hand base.
3. Choose parallels to set the workpiece off the bed so that the desired point of contact is in line with the clamp screw.
4. Slide the left-hand assembly to the right, almost contacting the workpiece; then tighten the base clamp bolts.
5. Tighten the clamp screw to secure the workpiece (but—caution—*overtightening* can bend the mill table).

There are two-part vises out there using various means to control tilt, but they may not work reliably. You *cannot* use dowel pins sliding in nicely reamed holes; they are almost guaranteed to bind. Instead of dowel pins this design uses shoulder screws in generous clearance holes, so the binding problem doesn't arise even if the workpiece is slightly trapezoidal. Strong compression springs hold the clamp plate to the left-hand base, which slides on Delrin or brass plugs.

10-3 ANGLE IRON BASES

Home shops may have a tough time cutting the 1/2"-thick angle. With a small bandsaw, this will take a while, so you might want to spend a few extra dollars having it roughed out by the steel supplier. Most of the dimensions are not important, so leave yourself the minimum of material to remove by machining.

The angle iron surfaces need not be machined, including the ones that rest on the mill table. Instead of machining, you can descale the surfaces with a carbide scraper, followed by P150 sandpaper (use spray adhesive to stick this on a scrap marble tile; then hold the base as if it were a sanding block—sanding any other way will give you a non-flat surface). For general clean-up, use a hand-held grinder with a Scotch-Brite abrasive disc.

10-4 MOUNTING BOLTS

Drill two clearance holes in *each base* for the mounting bolts (bolt size depends on the T-slot width). On the machine for which this vise was built, the T-slot width was 5/8", so the bolts were 1/2". To attach the keys, drill an 8-32 clearance hole on each base, centered *as accurately as you can* between the mounting bolt holes. If appearance is important, counterbore the hole (socket head cap screw suggested).

Figure 10-4 (see Section 10-7, below) shows the angle iron fillet milled away to create a pocket for the inner dowel pin. The intention here is to separate the dowel pins as much as possible, thus reducing the potential for angular misalignment—not strictly necessary, but a good idea. Use a center-cutting end mill on the same Y axis setting as the 8-32 clearance hole.

10-5 MILLING THE EDGES

If your interest is function rather than form, the edges of the angle iron can be left as they came from the saw. If you prefer something a little less ugly, end-mill the edges using a 1/2" or larger cutter. For another refinement, you can miter the corners with a bandsaw, then file or grind the edges smooth.

10-6 T-SLOT KEYS

The 3/8"-thick T-slot keys attach permanently to the underside of each base to ensure that the bases sit at right angles to the mill table every time they are installed. The length of the keys is not important—though the longer the better. The width of the keys is important, so expect a number of iterations when sizing them on the mill. Also expect some delicate filing in the final stage. Two T-slot keys are required.

The T-slots on your mill may vary in width by as much as ± 0.001" at various points. This means the best you can do is aim for a push fit that's not quite tap-in tight—but with zero wiggle room. The keys will be located by two 1/8" dowel pins and secured by an 8-32 socket head screw.

At this stage, drill and tap *only the 8-32 hole*, exactly on center; then set the keys aside.

10-7 JAW PLATES

Using your regular mill vise, skim-cut the faces of both plates to remove surface imperfections (the top edges will be finished in the final stage of the project).

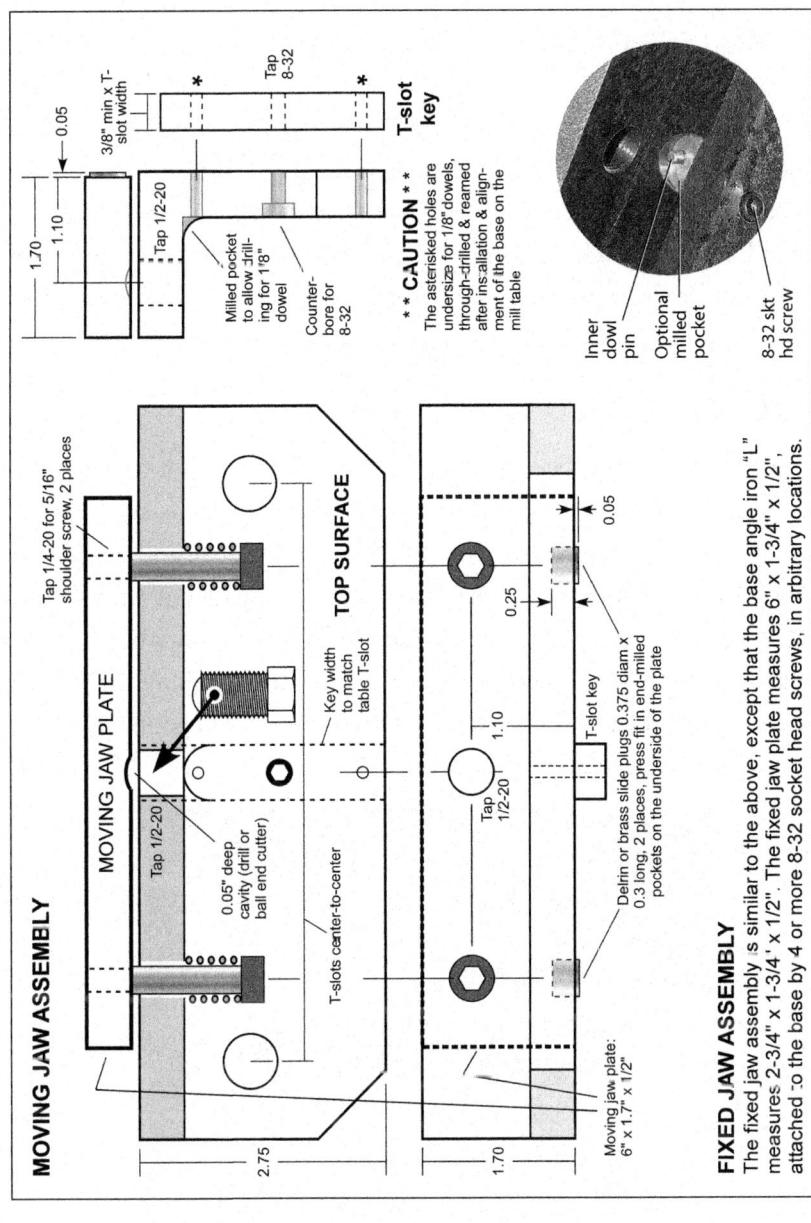

MOVING JAW ASSEMBLY

Tap 1/4-20 for 5/16" shoulder screw, 2 places

MOVING JAW PLATE

0.05" deep cavity (drill or ball end cutter)

Tap 1/2-20

Key width to match table T-slot

T-slots center-to-center

TOP SURFACE

Moving jaw plate: 6" x 1.7" x 1/2"

Defrin or brass slide plugs 0.375 diam x 0.3 long, 2 places, press fit in end-milled pockets on the underside of the plate

T-slot key

Tap 1/2-20

1.10

0.25

0.05

2.75

1.70

3/8" min x T-slot width

Tap 8-32

T-slot key

Tap 1/2-20

Milled pocket to allow drilling for 1/8" dowel

Counterbore for 8-32

0.05

1.70

1.10

* * **CAUTION** * *
The asterisked holes are undersize for 1/8" dowels, through-drilled & reamed after installation & alignment of the base on the mill table

Inner dowl pin

Optional milled pocket

8-32 skt hd screw

FIXED JAW ASSEMBLY

The fixed jaw assembly is similar to the above, except that the base angle iron "L" measures 2-3/4" x 1-3/4" x 1/2". The fixed jaw plate measures 6" x 1-3/4" x 1/2", attached to the base by 4 or more 8-32 socket head screws, in arbitrary locations.

FIGURE 10-4 Two-part vise details.

10-8 MOVING JAW PLATE

Machine the moving jaw plate to size, 6" x 1.7". (Compare this with the fixed plate's 1.75"—the difference is made up by plastic or brass plugs that slide on the mill table.) For the 0.05"-deep clamp screw cavity in the jaw plate, use a 1/2" ball-end cutter if you have one; if not, make a shallow conical cavity with a regular 3/8" drill. Drill and tap for the 5/16" shoulder screws; then drill two clearance holes (say, 11/32") through the face of the left-hand base to match, bearing in mind that the vertical distance from the mill table to the shoulder screw holes is 1.15". For the clamp screw, drill and tap a center hole (1/2-20 in preference to 1/2-13).

10-9 FIXED JAW PLATE

Machine the fixed jaw plate to size, 6" x 1.75"; then clamp it to the right-hand base as in Figure 10-5, making sure its bottom edge is aligned with the underside of the base. Drill through both the jaw plate and the base in four or six places using the tap drill size for your attachment screws, 8-32 or 10-32 socket heads. Now drill clearance holes and counterbores in the jaw plate using the same set of coordinates. With the assembly still held in the vise, tap the holes in the base, installing screws as you go to make sure the plate doesn't move in the process.

FIGURE 10-5 Attaching the fixed jaw plate.

10-10 CLAMP SCREW

The clamp screw is turned from 9/16" or 5/8" hex stock (Figure 10-6). The threaded portion should be about 3/4" long. If you have a ball turning attachment on the lathe, use it to cut the 1/2"-diameter nose; if not, file the stock to an approximately spherical shape. While at the lathe, don't forget the brass or plastic "slide plugs" for the bottom edge of the moving jaw.

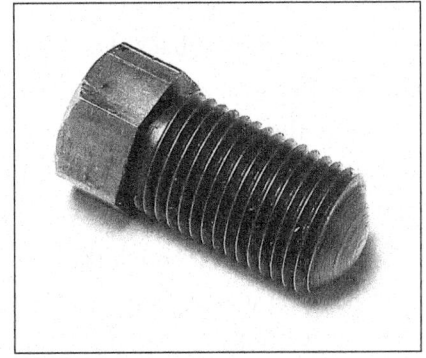

FIGURE 10-6   Clamp screw.

10-11 ALIGNING THE FIXED JAW

Install the fixed jaw plate on the right-hand base, fully tightening the attachment screws. Fit one of the T-slot keys on the base, fully tightening the 8-32 screw, shown in Figure 10-7 (1). Install the base in any desired

FIGURE 10-7   Front-to-back alignment. The 8-32 socket head screw securing the vise key is indicated by (1). The outer dowel pin location is indicated by (2).

location on the mill table. With the two mounting bolts in place, we come to the most important part of the project—aligning the fixed base with the Y axis, front to back. For this, you will need a dial indicator held in the drill chuck.

With the mill spindle locked (or set for lowest speed if not lockable), fully tighten the front bolt on the base. Partly tighten the other bolt. Starting at the tighter end, preload the indicator 0.05" or so; then note its reading as it traverses to the far end. If the difference from one end to the other is, say, 0.02", correct it by tapping the far end of the base to halve the difference, 0.01". Tighten the far-end bolt a little more; then return the indicator to the tighter end and repeat. Always work from the tighter end. Done right, it can be lined up within ± 0.001" in two or three passes, assuming the fixed jaw plate is truly flat. Fully tighten both bolts; then check again—this is a one-time opportunity to get it right.

10-12 DRILLING THE VISE KEY FOR THE DOWEL PINS

Ideally, the 1/8" pins used to hold the bases firmly aligned would be hardened dowel pins in reamed holes. If dowel pins are not available, use 1" lengths of 1/8" drill rod. You may not need to ream if your 1/8" drill is sharp (secure the pins with Loctite if necessary. Be sure the fixed jaw is perfectly aligned; referring to Figure 10-4, dril through the base and vise key for the pins on the same centerline as the 8-32 screw. Place the inner and outer holes as far apart as possible. Remove the assembly from the mill to install the locating pins.

10-13 FINISHING THE MOVING JAW

Assemble the moving jaw to the left-hand base using shoulder screws and compression springs. As designed, the shoulder screws were 1-1/4" long, and the springs 1-1/2" long, uncompressed. The choice of spring is fairly arbitrary, but aim for a matched pair with enough force to give a solid feel to the assembly.

Insert the "slide plugs" into the holes on the bottom edge of the moving jaw. Install the clamp screw; check its pushing action, then back it off completely to fully seat the jaw plate firmly against the base. Fit the second vise key to the moving jaw base, fully tightening the 8-32 screw. Install the left-hand assembly on the mill table so that the moving and fixed jaws touch firmly. (This ensures proper alignment of the left-hand base without indicating, as was done for the fixed jaw in Figure 10-7.) Tighten the left-hand mounting bolts and then drill 1/8" holes through the base and key as described in Section 10-12.

Finally, reassemble the two halves of the vise; then take an equalizing cut over the top edges of both jaws and both bases (Figure 10-8).

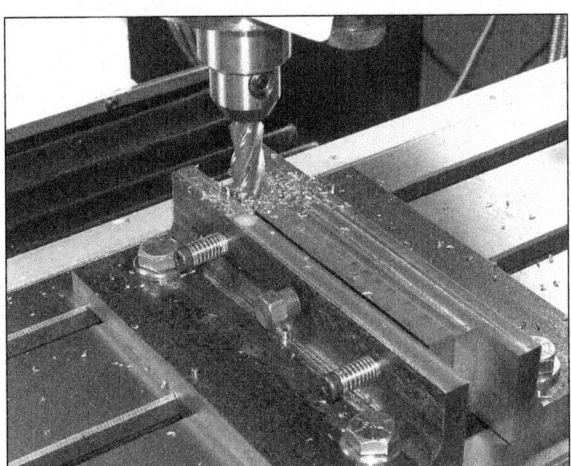

FIGURE 10-8 Finishing the top edges of the jaws and bases.

Index

About the Author

Richard Rex has worked on lathes and milling machines since his teen-years in a home shop, and later on a variety of production machines (his current home shop setup has a 12" x 36" lathe and a Bridgeport mill). More recently, he has set up several engineering lab model shops from scratch, with the usual complement of Hardinge lathes and Bridgeport mills.

Richard worked for 10 years in product marketing management with Hewlett Packard and Brown Boveri in the UK. In the US, he has been CEO of several engineering/manufacturing companies. He has been a "tech writer" throughout his working career, starting with monographs on signal processing with HP, later (as CEO) data sheets and application notes for a wide range of his company's products. In recent years he has written and illustrated 30+ manuals and tech bulletins for a machine tool distributor in Pittsburgh.